MW00759889

HISTORY AND THE SEA

HISTORY AND THE SEA:

Essays on Maritime Strategies

by Clark G. Reynolds

University of South Carolina Press

"America as a Thalassocracy—An Overview," previously published as "The Sea in the Making of America," Copyright © 1976 U.S. Naval Institute.

"Reconsidering American Strategic History and Doctrines,"
previously published as "American Strategic History and Doctrines: A Reconsideration,":
Copyright 1975 by the American Military Institute

"The Continental State Upon the Sea: Imperial Japan,"
previously published as "The Continental Strategy of
Imperial Japan," Copyright © 1983 U.S. Naval Institute

"Ernest J. King and American Maritime Strategy in the Pacific War," previously published as "Admiral Ernest J. King and the Strategy for Victory in the Pacific," in the *Naval War College Review*.

"Uses and Misuses of World War II Maritime Strategy Today," previously published as "The Maritime Strategy of World War II: Some Implications?" in the *Naval War College Review*.

Published in Columbia, South Carolina, by the
University of South Carolina Press

Manufactured in the United States of America

Library of Congress Cataloging-in-Publication Data

Reynolds, Clark G.
    History and the sea: essays on maritime strategies/by Clark G. Reynolds.
        p.   cm.
    Bibliography: p.
    Includes index.
    ISBN 0-87249-614-7
    1. Sea-power–History. 2. Naval history, Modern. 3. Naval history.
    4. Naval strategy–History. I. Title.
V25.R48   1989
359'.03–dc19

        88-34604
        CIP

CONTENTS

PREFACE

The strategic uses of the world's oceans by the great powers of the past have long interested scholars, makers of foreign and defense policies, and the practitioners of such policies—generals and admirals. Whereas such students have traditionally focused on individual nations, however, the present volume defines patterns of strategic behavior by *types* of nations: maritime and continental. In so doing, it suggests fresh historical interpretations of how the sea has been utilized by the great powers in general but especially by the United States, its allies, and its adversaries.

The first three chapters define maritime strategy and its applications by the great sea-oriented empires or "thalassocracies" of history. Chapter 1 is a broad historical and geographical overview of nations upon the sea in peacetime and war; it sets forth the proposition that only particular strategic options have been available to maritime and continental great powers. Chapter 2 considers in detail the idea that thalassocracy has been a historical "force" from ancient times to the present. Chapter 3 criticizes the great savant of "sea power," Alfred Thayer Mahan, as having been heavily deterministic in his view and applications of history.

The United States is examined as a genuine thalassocracy in Chapter 4, an introductory explanation, and in Chapter 5, an in-depth, heavily annotated historiographical reinterpretation of American strategic behavior since colonial times. Inasmuch as the United States reached full strategic maturity in World War II, the next three chapters focus on specific aspects of that conflict: Japan as a continental rather than maritime empire, and the strategic genius revealed in the persons of two Americans—Admiral Ernest J. King and General Douglas MacArthur, the latter also for the postwar period.

Finally, contemporary maritime strategy is treated through the historical dimension. Chapter 9 interprets contemporary American options in light of the World War II example. Chapter 10 explores Russia's activities upon the sea over the sweep of eight centuries as a major factor behind current Soviet strategy.

This book is the result of my thinking over a quarter-century about the interrelationships of the sea, nationhood, and grand strategy throughout the past. The first nine chapters appeared as essays in professional history and defense publications between 1972 and 1986; the last one, on Russia, appears here for the first time.

Clark G. Reynolds

HISTORY AND THE SEA

1

STRATEGIC USES OF THE SEA —PATTERNS AND OPTIONS

———

The special language of seafarers often deters "landlubbers" from understanding the particular ways in which the sea has been utilized by nations throughout history. Ships are simply ships—men-of-war or merchantmen—to most statesmen, generals, historians, and even the lay public, who consequently lack an appreciation of the subtle applications of what Mahan termed "sea power" to national goals.

This chapter therefore introduces the geographical dimensions of the sea in the past, the distinctive natures of the great maritime and continental empires of history, the theoretical schools of historical and material strategy, and the separate applications of naval power by maritime and continental empires and by small nations. Elementary though this discussion may appear, it is basically original, synthesizing most past thinking about the strategic uses of the sea but discerning certain patterns of strategic behavior. And it is fundamental to understanding fully the rest of the book.

The chapter originally comprised "Book One: Command of the Sea—and the Alternatives" in *Command of the Sea: The History and Strategy of Maritime Empires,* published as one volume by William Morrow of New York in 1974 and by Robert Hale of London in 1976 and revised as two volumes by Robert Krieger of Malabar, Florida, in 1983.

1

History and the Sea

. . . a vague feeling of contempt for the past, supposed to be obsolete, combines with natural indolence to blind men even to those permanent strategic lessons which lie close to the surface of naval history.

—ALFRED THAYER MAHAN

The sea covers 71 percent of the earth's surface; the remaining 29 percent is mostly land mass, generally occupied by human groupings in areas readily accessible either to the great saltwater oceans or to the many inland fresh-waterways, the major lakes and rivers. The sea-borne communications between these settlements have often been a dominant force in the history of great nations. Thus, command of the sea and inland waterways has remained a key political and strategic concern of seagoing peoples throughout history.

Geography is the major determining factor in any nation's ability to utilize the sea commercially and to defend its political and economic integrity from overseas attack. Thus, each nation tends to orient its political, economic and military life around the advantages of its geographical position vis-à-vis other nations. And history reveals that this orientation has usually favored either the ocean-maritime element or the continental. No nation has yet been able to afford the sheer expense of sustaining both a large army to control its continental frontiers and a large navy to maintain control over vast areas of water.

While immediately adjacent waters have been of constant concern to every nation, the specific waterways politically disputed throughout history have changed in emphasis with the growth and reach of Western civilization.

In the ancient and medieval periods of history, the Mediterranean Sea enclosed the areas of major political dispute. These were the Eastern Mediterranean, from the Aegean Sea and the Dardanelles in the north, to the islands of Cyprus and Crete in the center, to the Phoenician coast and Nile Delta in the south; the Central Mediterranean, from the Adriatic Sea and Italy in the north, to the Ionian and Tyrrhenian seas and Sicily in the center, to Carthage and Malta in the south; and the Western Mediterranean, the clockwise circle of southern France, Corsica, Sardinia, North Africa, Gibraltar, and Balearic Islands (Majorca and Minorca) and the Iberian peninsula.

Strategic Uses of the Sea

In the modern period, roughly 1500 to 1900, maritime relations between nations were geographically governed by the so-called funnel of Europe, the sea lanes of the North Atlantic Ocean that passed through the narrow Strait of Dover and English Channel, connecting the North Sea nations with France, Spain and Portugal to the south. At the focus of this activity lay England, around which the major maritime events of this period revolved. Flanking this center of maritime activity and international rivalry were three virtual overseas lakes, all linked to the ever-lengthening European trade routes: the Baltic Sea to the north, the Mediterranean to the south and the Caribbean to the west. Secondary areas in this period were the South Atlantic, Indian and Pacific oceans, and the Black Sea.

In the twentieth century, the rise of non-European powers has made the two great oceans, the Atlantic and Pacific, of equal importance. All the other oceans, seas, lakes and rivers are secondary in importance, although it must be recognized that international rivalry upon the seas has become global in this period.

Throughout history, the chances of a nation becoming a dominant power upon the sea, with a large merchant marine, overseas colonies and a strong navy to protect both, has been determined always by favorable geographic conditions. Insularity is of prime importance. Without land frontiers to defend, a maritime nation may minimize its army for home defense and simultaneously be able to project its commercial and naval strength overseas. Placed in the dominant geographic position relative to rival land powers, oceanic states such as ancient Athens, modern Great Britain or the contemporary United States have been able to emerge as dominant maritime nations. And their strategies have rested upon their ability to command the sea.

Nature must be kind to aspiring maritime nations by also providing them with many good bays and inlets, protected from storms and extreme tides, adverse winds and weather, yet deep enough and large enough to harbor many large ships. In the ancient period, smooth coastal waters were essential for the operation of rowed galleys between these harbors. In the modern period, favorable prevailing winds were essential to propel sailing ships from these ports. In addition, long and wide rivers with sizable outlets, deltas or bays opened oceanic trade and naval operations to and from the interior.

Without these strategic geographical assets, no nation has ever been able to achieve a lasting political and economic dominion over the seas. Continental powers so limited have had to adopt special measures or alternatives to compensate for their inherent geographical shortcomings. As a result, two separate strategic policies recur throughout history as great maritime and continental powers have confronted one another.

EMPIRES, NAVIES AND STRATEGY

Great powers have evolved to the height of their political prestige and might by becoming imperial nations, though they may not choose to accept this label. Though the meaning is forever debated among historians, *empire* may be generally defined as the supreme authority of a large and powerful nation over considerable territory beyond its immediate borders. *Imperialism* is the policy of extending that authority further, by acquiring colonies or dependencies. Though any sprawling nation may be (and many have been) considered *imperial*, common usage throughout history has tended to regard mostly those nations with overseas holdings as truly deserving of the label *imperial*. Nevertheless, landed rulers have often preferred the grandiose title of *emperor*, and occasionally they have deserved it in the political context of having far-flung subject peoples. But for purposes of convenience, historical accuracy and strategic analysis, real imperial powers may be considered as those great nations in history which have based their national political and economic policies and strategies chiefly on *maritime* activities: commercial trade, overseas possessions or dependencies, and naval forces. The other great powers, non-maritime by nature, may be simply labeled as continental, relying chiefly upon the produce and manufactures of the national homeland and whatever political or economic advantage it can gain from its continental neighbors. Lesser nations may vary in their orientations between livelihood on the sea and the soil, but their larger destinies are usually subject to the actions of the great *maritime/imperial* and/or *continental* powers.

Strategic Uses of the Sea

Maritime powers, vulnerable to external pressures on their food supply, raw materials and power sources and thus primarily interested in maintaining their economic wealth through overseas trade, have therefore sought to enforce a reasonable state of international order on the high seas, so that the economic lifeblood of their merchant economy should not be interrupted or threatened. They have consequently depended upon their navies to maintain that order by policing both the trade routes to their overseas markets and also the oceans of the world wherever the merchant vessels ply. The national interest of maritime powers has generally dictated a policy of either monopoly or the free use of the seas, whichever best profited their own economies. In addition, uninterrupted sea-borne communication among such a maritime people has tended to bind them politically and culturally as well as economically.

Domination of the seas by a great maritime power in the cause of economic and thus political stability has resulted in protracted periods of seeming "peace." Each so-called *Pax—Romana, Britannica* and *Americana*—has really been naval peace, where supremacy at sea provides a major deterrent against serious challenge by unfriendly opponents. In reality, *pax* or peace has been a misnomer, as true peace can exist only within a political vacuum. And political vacuums are— sad to say for "peace"-loving peoples—a virtual impossibilty. Rather, periods of international stability and political orderliness are made possible by a precarious balance of tensions between two or more great powers. That the prolonged maintenance of such balances is difficult is evident in the multitude of "policing actions" and wars fought in the name of maintaining some balance of the great powers.

If one needs a working definition of the balance of power, it is the distribution of nearly equal political-military power between two competing nations or groupings of nations so that normal economic intercourse remains unrestrained—or, so that relations between nations remain "peaceful."

Maritime empires have been expensive to protect. A large navy is necessary to patrol the colonies and trade routes of the empire and, if necessary, the seas beyond them. Also required are police-type (pacification) ground forces to enforce internal order within the overseas possessions. But the biggest expense is ships, which are costly to build, arm, supply, man, keep up, repair and eventually replace. Still, the investment in a navy is an investment in political and economic security. Ignorance of this fact has led to strong internal political opposition to large naval budgets in all nations having warships. So necessary are the navy and constabulary troops to the welfare of a maritime empire that one may generalize that the stability, loyalty and trustworthiness of overseas colonies and dependencies are directly proportional to the strength of the mother navy.

Just as the navy helps to determine the political and economic destiny of great maritime nations, so too is it a dynamic force in the cultural and social aspects of the national life, for maritime nations or thalassocracies have two important social advantages over continental powers, stemming from the natural accident of their geographic location.

First, since maritime nations are usually insular, they enjoy what might be called national privacy. With no unfriendly powers poised on their borders, they enjoy something which continental peoples have always considered a luxury—no large standing army or national psychosis of impending attack. If maritime peoples can rely upon a formidable navy operating literally out of sight of their homeland to insure their insularity, they can ignore such culturally inhibiting forces as military frontiers and forts, military strongmen and despots, standing alliances, frequent invasions and wars, and the whole mosaic of problems involved in counterattacking, occupying, defeating and reconstructing an enemy nation. Such countries have had the advantage of *time* over their constantly embattled peers on the continents, time in which to develop their institutions and industries in relative peace. Maritime insularity, then, has been a key ingredient in intellectual ferment, the growth of applied technology, and the fostering of democracy. Isolated and thereby well-defended men have tended to be free men, free to think and to apply their ideas to machinery and to government.

The other advantage of the maritime nations over the continental, and closely related to this tendency of isolated peoples to promote free thought, is that by the very nature of their economic life such nations have placed a high premium on the worth and skills of the individual. Merchant traders on land or sea are an independent lot anyway, yet the growth of the merchant class has been more rapid and pronounced in countries that depend on overseas trade. Continental states have taken decades longer than their maritime counterparts to bridge or close the great gap between aristocracy and peasant class. Equally if not more significant has been the individual sailor. Life at sea is a high adventure, involving the wisdom and raw stamina of men against not only an enemy in battle but against nature. The spirit of great maritime peoples has been embodied in their naval heroes and their tireless explorers, who in the past conquered unknown frontiers overseas and in the contemporary period have been mastering the poles, the ocean depths, the skies and space beyond.

Navies especially have been a bedrock of individualism, while the naval profession nurtures this individualism by demanding the very qualities that shape great peoples—discipline, creativity and a high degree of practical intelligence. Discipline at sea is more than abiding by the orders issued from a rigid hierarchy of authority; it is also the self-discipline of the individual seaman in the face of constant danger. Practical common sense and the ability to improvise when short on doctrine or material are equally essential for survival at sea. Rugged and outspoken when in their own element, sailors have always possessed a large measure of tolerance and yet a keen sense of justice and fair play. They respect authority, but can be pushed to rebel against unreasonable commanders.

This thalassocratic individualism has remained an essential ingredient not only of effective navies but of great maritime nations. Small wonder, then, that such nations place such value on their seamen, merchant as well as naval. At one with the sea, which he must eternally battle to control in the struggle between man and the elements, the individual sailor develops a self-confidence and pride of service seldom equaled in armies. Relatively unaffected by shifting political winds ashore, the sailor fashions a hardened sense of duty and loyalty to his ship and his profession. Hence navies have usually been among the least political and most stable of institutions in maritime nations. A vital component of the democratic spirit, navies have remained a pillar in the support of free institutions, socially as well as strategically.

History and the Sea

Politically, however, navies have always been weak in asserting themselves overtly. Sailors learn their professional skills at sea among a small ship's company, where administrative and political considerations are minimal. They are technical experts, skilled in the technology of service at sea and sensitive to the inherent fragility of their machines. And seamanship knows no politics. By contrast, land-oriented officers deal with vast administrative organizations of many men and large tracts of territory and are in constant physical association with the political organs of government. These army men are the most skilled administrators in any military operation involving both land and sea forces.

Naval officers have therefore lacked the political polish of the generals and have tended to remain aloof from politics or to adopt a seemingly safer, conservative approach. Even in political revolutions, navies tend to remain as nonparticipants. Broadly speaking, sea power works slowly and subtly, whereas generals, politicians and the people at large are impatient for direct, immediately apparent results, as with armies on the march. These elements of society therefore tend to view expensive navies with suspicion when confronted with often short-term or superficially more pressing domestic and diplomatic concerns. Resisting "liberals" out to cut the fleet, admirals tend to oppose changes to their service from which recovery will be difficult when the national mood again suddenly reverts to a frantic awareness of actual naval requirements. Thus sensitive to criticism by politicians and their constituents, flag officers usually fare rather badly when embroiled in broad political disputes.

Strategically, then, as well as politically, navies and their admirals tend to favor the *status quo*. They need *time* for their exercise of command over the seas to be felt, and they require an economically superior government to support their own needs for a superior fleet. Unlike armies, whose strength is built on manpower, navies depend more on the technology of their ships and weapons, which are infinitely more costly than the raising of armies. Furthermore, navies are intolerant of forces aimed at disrupting the order they enforce. Consequently, they have tended to regard interlopers into their imperial system as illegal enemies. Most outstanding of their "outlaw" foes throughout history have been rebels, smugglers and pirates.

Strategic Uses of the Sea

Colonial settlers or subject peoples sooner or later develop a desire to share with the mother country the fruits of their labors. But if such empires persist in exploiting these areas and not sharing their wealth with such indigenous folk, they will likely be faced with revolt. The navy is then charged with the task of reestablishing order in the colonies, an expensive task that requires considerable effort and diversion of naval material from other pressing strategic activities. The rebels must be isolated from outside help, thus requiring warships to police the seas in the troubled area. Imperial troops must be sealifted to the place, then supported logistically by the navy and merchant marine. And once suppressed, the rebellious possession must then be closely watched by a garrison and the navy, which is dependent upon the area for its advanced base of operations. Thus navies, like their governments, regard any political upheaval as dangerous to imperial stability. Rebels cannot be tolerated if order (or "peace") is to prevail.

Pirates have generally been considered to be outlaws preying upon merchant shipping and the colonies. In the early ancient and early modern periods, before the advent of maritime empires based upon far-flung political and economic order, piracy and privateering (commissioned private raiders by an enemy) were considered to be more or less respectable professions. As civilization became more rigid and erected legal safeguards to protect the economic wealth of the nation-state, the pirate appeared increasingly as a violator of civilized international law upon the high seas. A "barbarian" (i.e., uncivilized), this outlaw became such an outrage to imperial nations that he was effectively eliminated by them before the end of the modern period. In their role of policing the oceans of the realm, navies have always had to be able to suppress piracy effectively on the high seas. Without navies, pirates flourish.

Therefore, to maintain political and economic stability upon the oceans of the world and throughout their own empires, maritime nations have depended upon their navies. Their strategies have thus embodied the ability to deter rival powers from interfering with their own maritime activities, to suppress pirates and to police the trade routes and overseas possessions of their own empires. To do these things, such navies must be able to command the seas. Opponents of major maritime nations have been the competing maritime nations, major continental powers, or a combination of both.

Continental powers have been governed by very different geographical considerations, the result being that the naval or maritime aspects of these nations have generally been secondary in political and strategic importance. A systematic analysis of great land powers is not within the scope of this study, except as such nations have attempted to utilize waterborne economic power and military force in the face of strong maritime rivals. Generally speaking, continental powers have depended upon overland communications for their economic wealth and upon large armies and fortifications for their political and military security. Constantly exposed to and threatened by overland invasion, such nations have reduced their armies only at their peril. Agriculturally (and later industrially) based, these nations have depended mostly on manpower for defense—the mobilizing, disciplining and administration of large armies. The effect has been a general tendency toward authoritarian government, national regimentation and a servile population. In general, whereas the independent merchant class has typified maritime societies, a powerful landed aristocracy has dominated the life styles of continental powers until very recent times. Their political base has literally been the land. And their political goals have usually been obvious—defense against the invader—as opposed to nations which depend on more subtle goals to be gained through the application of sea power.

The attempt of continental powers to operate navies has consequently been frustrated not only by geographic limitations but by related political, cultural and social contradictions. Politically, the ruling landed aristocracy (or industrial managers in the contemporary period) is preoccupied with defending the *status quo* at home, preserving the government from internal upheaval and external attack. Such a class has little appreciation for expensive overseas enterprises and tends to be too rigid to adopt the techniques and innovations of the maritime powers. So it invests primarily in the army, subordinating the navy to these continental objectives. Culturally and socially, such nations do not generally enjoy the spirit of individualism engendered by maritime adventurers, but rely rather upon their people as a mass. When in fact such land powers have attempted to create a maritime empire replete with a navy, as some have, their merchant mariners usually have emerged profitably, but the navy soon discovered itself outside the mainstream of internal politics and national life. This apolitical trait is consistent with the naval politics of maritime nations, except that eventually the continental navy finds itself manipulated into virtual extinction by the dominant army-supported class.

History indicates that the most viable solution a continental power can seek in its quest for a naval presence is—in addition to its own small navy—alliance with a maritime nation. Depending upon such an ally has proved risky, for obvious reasons, but alliance has probably been the most workable compromise solution to meet an otherwise almost impossible need.

So, strategically there have been great maritime nations and great continental nations. One type has shaped its strategy largely around the overseas thrust of its navy. The other has depended strategically upon the defensive stance and occasional offensive thrust of the standing army (and later, land-based air forces) overland. Despite exceptions throughout history, these distinctions between land and sea powers have generally held true for major peoples and nations.

STRATEGIC HISTORY

The history and strategy of maritime empires have been shaped not only by geography and men but by naval technology as well. Indeed, the fact that a navy exists is a sure indication of civilization and its growing technology. A full understanding of the technological element is thus crucial, because the misunderstanding of dominant weapons and other technical aspects of defense policy has often led historians and strategic analysts alike astray from the essential lasting principles of maritime power.

More so than armies, navies have required many years to evolve, due to the technological nature of their relatively more sophisticated equipment. Ships must be designed to incorporate the latest innovations in naval architecture and weaponry, then constructed over a number of years—up to four years for the largest warships of the modern and contemporary periods. These latter, often called capital ships, have usually been the yardstick of naval power, the ship-type around which the tactics of a fleet are formulated. Essential to evaluating the actual strength of such vessels, however, is an understanding of their cruising characteristics, the propulsion system and operating range, main and secondary armaments, defensive protection, signals communications and control over the ship's operation. Smaller vessels deserve equal attention as seagoing machinery. And the whole weapon (or weapons system, in recent parlance) depends on the skill and well-being of its operators. The officers and crew fashion the merchant ship and the vessel of war, so that a given naval technology is only as good as the training, experience, clothing, feeding, health and morale of the seagoing technicians.

Furthermore, naval technology and weaponry are utterly useless if the techniques of employing them prove wanting. Unfortunately, the inferior employment of potentially superior weapons has been an all too frequently repeated mistake of maritime and continental powers alike throughout history. This has been due largely to the very human assumption that the dominant weapon should determine the strategy and tactics of a given period. Often this has proved to be a sound assumption. But just as often, strategic and tactical realities change, rendering the apparently dominant weapon less effective or even downright obsolete.

This difficult strategic problem, of weighing the weapons technology of a given period against historical experience, has no simple solution. Indeed, strategic thought has tended to polarize into two general schools in the industrial nations of the late modern and contemporary periods. They are the material and historical schools of strategic analysis.

The *material* school rests upon the assumption that the dominant military hardware or weapon—the material strength—at a given time creates such an overwhelming superiority that it alone generally satisfies the nation's defense needs. This line of thinking is usually concerned primarily with waging or deterring *total war* between superpowers. It further includes a recognition that a technological ceiling has been reached, creating not only superior weaponry but perhaps also national superiority in overall technology, political and economic systems and culture. Such a weapons superiority in the Western world has given certain nations the power to dictate the course of international affairs or to balance off the weaponry of an equally strong power in a strategic stalemate. In either case, such a nation assumes the position of controlling the balance of political power. In the industrial and scientific environment since the early nineteenth century, such technological determinism has tended to dominate strategic thinking.

The *historical* school of strategy rejects this determinism by examining the past conduct of competing nations in order to understand all historical forces at work and thus the various alternative approaches to strategic problems. Along with the problems of waging or deterring total wars, historical strategists are also concerned with *limited war*, with the diplomatic and legal aspects and alternatives to conflict, and with the problems of combating primitive or undeveloped peoples who do not honor the technological assumptions of advanced Western nations.

In the search for simple solutions and panaceas in strategy, both schools of analysis have erred in overstating their respective cases. Many material strategists have viewed superweapons as a panacea, while strategic historians have often expected history to repeat itself, thus committing the folly of depending completely on the proverbial dead hand of precedent. The overconfidence of both groups has frequently led to unfortunate consequences throughout history.

The difficulty of adequately combining the military principles and ideas of both history and advanced technology in order to formulate strategy has been due partly to the sheer chance of historical timing. Until the mid-nineteenth century, each nation-state had been generally governed by the same constant factors—agricultural wealth, land-based aristocratic political institutions, and the extent of territory and raw materials under its control. The generally constant level of technology tended to limit the size of armies and navies, therefore making the strategic options open to great maritime and continental powers fairly predictable. Indeed, by the 1880s, the Old World seemed to have reached a state of eternal peace, or at least a political stability in which only limited wars could occur. At that time, however, a number of brilliant maritime-oriented strategic historians emerged to examine systematically the forces that had shaped their world. The leaders of this intellectual ferment were the American Mahan and the Britons Corbett and Richmond. Their penetrating questions, ideas and writings epitomized the historical school of strategy.

However, the pinnacle reached by the Old World and its strategic analysts was accompanied by the end of the modern period. With the advent of the twentieth century came advanced scientific thought, systematically applied to the new weapons of unprecedented power. In 1914, the Old World figuratively vanished as these weapons of the new technology were unleashed on the battlefields of Europe. Limited war and the relative peace also seemed to disappear in the new era of total technological war. So blinding was the new technology in its destructiveness that the Old World and its historical lessons were all but forgotten. In their place arose the material strategists such as the Italian Douhet, the Briton Trenchard and the American Mitchell who envisioned military success in the aerial superweapons of mass destruction. The events of this contemporary era, with its world wars, airborne nuclear weapons and resultant technological determinism, have dominated strategic thinking to the present. In fact, though, the efficacy of this school was finally revealed as wholly inadequate by the Cuban missile confrontation of 1962.

Indeed, by the 1970s, mankind had certainly reached a major historical watershed in all its activities—political relationships, communications, social habits, medicine, the need for control of unchecked technology, population and pollution—and in sheer scientific advancement, symbolized most dramatically by the landings on the Moon. Surely the time has come for new hypotheses and a fresh synthesis in strategic thinking as in the other aspects of human relations.

STRATEGIC APPLICATIONS OF NAVAL POWER

The examination of the strategic history of navies and maritime empires requires not only the analysis of evolving strategic principles of naval warfare but of the long and often slow development of tactics, logistics (supply), command and administrative control, communications, ships, weapons and other aspects of naval life and technology. Through studying the flow of history and navies in history, it is possible to discover both the impact of sea power upon history and the impact of historical forces upon navies, empires and strategy. Finally, along with the successful application of the strategic doctrine of command of the sea, an appreciation is possible of the strategic alternatives to such a policy.

As guidelines to this examination, a list of theoretical strategic applications of naval power is useful. Since no one type of nation has existed throughout history, however, there has been no one type of navy or naval strategy. The guidelines of applied naval power, there-fore, are aimed at understanding three different types of navies, that of maritime or "blue-water" nations, that of continental nations and that of small nations.

For *maritime* nations, the navy has been the main strategic arm of the nation's defensive structure, dominating the defensive policies of the home government, maintaining a generally *offensive* stance, and operating mainly on the "blue water" of the high seas. The army of such a nation is usually small by contrast, so that for large-scale land operations, the maritime nation usually must depend upon a large continental ally. This navy has several functions, all of them geared to the principle of achieving *command of the sea:*

1. Maintain a superior fighting fleet either (a) to seize command of the sea, or (b) to deter an enemy from attempting to control the sea. In wartime, this fleet is used as the *active* force to seize, exercise and maintain control over disputed waters. The waters in question are usually the open ocean, but may also include coastal areas, lakes and rivers. In periods when no declared war exists, this fleet acts as a *passive* force by demonstrating to competitor nations that it has the ability to dominate the seas; in this way, it deters aggression by its threat of seaborne retaliation.

2. Defend against invasion. A defensive requirement, this task calls upon the fleet to protect the shores of the home country either by destroying or otherwise neutralizing the enemy fleet in wartime or in "peacetime" by threatening a competitor navy with destruction.

3. Protect maritime commerce. Also a defensive need, this re-quires the fleet to keep open its own sea lanes for its merchant ships. It

may utilize overseas bases and its own mobility either to escort merchant vessels or to clear the seas of enemy raiders, pirates or other interlopers.

4. Blockade the enemy coast. An offensive requirement of the fleet, the seas around the enemy coast must be denied the enemy for the use of his merchant marine, for neutral vessels trading with him and for his own vessels of war. As long as the enemy fleet survives, the blockade is generally *naval;* after the enemy fleet is destroyed or otherwise neutralized, the blockade is primarily *commercial,* aimed at stopping enemy trade. In either case, it may be a *direct* blockade, with the fleet actually remaining on station off the enemy ports, or it may be *indirect,* the fleet observing and thwarting enemy ship movements from a considerable distance away.

5. Engage in combined operations. Either in offensive or defensive situations, the blue-water fleet must be able to sealift ground forces, army and marines, to and from a disputed area, the goal being invasion and capture of an enemy's overseas possessions and bases. The fleet must be ready and able to invade the enemy's home country, in the event that a successful commercial blockade does not compel the enemy to submit. In all such amphibious landings, the fleet provides tactical bombardment to cover the assault, then logistical support of the beachhead. In such *sealift* and *support* operations, the navy cooperates with the ground forces (which in recent times have included land-based air forces) by keeping open their lines of communications, by policing coastal and inland waters, and by commanding the sea (and the air, in recent times).

6. Provide strategic bombardment. The ultimate expression of naval superiority comes when naval power can be projected inland against the vitals of the enemy homeland. This function is not always required, as an enemy may sue for peace or may surrender to the naval and commercial blockade first, or the army (and strategic air forces) may be better equipped for this task, which belongs essentially to a continental strategy. Nevertheless, recent technology has given the blue-water navy the capability to project its firepower well beyond the enemy coastline.

For *continental* powers, the army (and lately, in combination with the land-based air force) has been the main strategic arm of the nation's defense. For blue-water operations of a broad offensive nature, this nation will best rely upon an allied maritime power. Its own navy usually maintains a *defensive* strategic stance, governing its operations to enhance the strategic advantages of the army. This navy has several functions, all dictated by strategic needs on the continent:

1. Defend against invasion. This navy must augment coastal defenses to help repel an enemy fleet from the continental periphery, adjacent lakes and rivers, and overseas possessions for the purpose of not allowing the enemy to establish a bridgehead on the coast for an invasion or for small-scale raids.

2. Engage in combined operations. In *support* of the main strategic arm, the army, the continental navy may gain command of the sea by default, that is, by the combined effort of army and navy in capturing enemy ports mainly by overland attack, thus depriving the enemy fleet of its crucial bases. To do this, this navy should have a limited *sealift* capability, for transporting troops over short stretches of local waters. The magnitude of such offensive amphibious operations may vary from small raids to an actual mass expeditionary force invasion. But a major difficulty of such operations is that a continental navy usually lacks both command of the seas to carry out the assault and a sophisticated amphibious doctrine due to its lack of experience. The continental navy may also provide a complementary offensive capability by combining, in recent times, with the army or air force to project its firepower into the enemy interior. But such strategic bombardment remains under the control of the senior services, as it is still part of the continental strategy.

3. Attack enemy commerce. Using the technique of what the French call *guerre de course*, the continental navy operates small squadrons or single ship units to prey upon enemy commerce. If utilized in overwhelming strength, this offensive function may assume the proportions of an effective commercial *counterblockade*, preventing vital war supplies from reaching the maritime enemy's homeland.

4. Maintain an efficient second-class fighting fleet either (a) to restrict enemy offensive action, or (b) to deter an enemy from attempting to dominate local waters. These closely related objectives may be achieved by the construction of a force of sophisticated naval

vessels, ship-for-ship at least slightly superior to their individual counterparts in the enemy navy. Such excellent warships deploying singly for sporadic operations can force their maritime adversary to deploy a significant number of his own ships to deal with them. Or such superior single units can be combined to present the appearance of a formidable fleet capable of blue-water operations. Loosely defined as a *fleet-in-being,* it can deter a maritime power from aggressive action, or in actual war it can tie down the enemy blue-water fleet from other vital pursuits in order to keep track of its movements. (Correctly, however, a fleet-in-being seeks to hold the defensive until it is able to assume the offensive, a concept practiced more by hard-pressed maritime nations than continental ones.) A well-handled continental navy of superior vessels, though inferior quantitatively, can thus have a pronounced effect on restricting the actions of a blue-water navy against the continent. In the contemporary period, the possible use of such vessels for projecting their firepower strategically into the enemy homeland further increases their prestige in any continental strategy.

Thus, if such an effective continental navy is directed by enlightened leaders in the government and is blessed with a generally bungling blue-water adversary, it can win command of the sea vicariously. If such conditions are just right, it can in some measure neutralize the enemy fleet by keeping it off balance, cutting supply lanes to the enemy homeland with a counterblockade, and thus protecting its own coast and merchant marine. It can sealift ground forces over limited distances and possibly provide some measure of tactical and strategical bombardment. But as long as the enemy blue-water fleet exists in any real strength, the continental navy can never maintain control over the blue-waterways. Whatever brief command of the sea it may enjoy, that command is only temporary and must be exploited quickly in order to serve the ends of the continental strategy.

For *small* powers, armies and navies alike can usually only hope for major success by allying with a great continental or maritime power and adopting its particular strategy—that is to say, unless they are fighting an equally minor power, in which case their strategy is dictated by the strengths and weaknesses of their adversary, emphasizing appropriate ground or naval forces. In any case, the minor navy must concentrate on three immediate tasks:

1. Defend against invasion. Inshore craft and naval weapons can be used to augment the national army and allied navy to help thwart an overseas or overland enemy attack.

2. Police local waters. Inshore and river craft and occasional large cruisers are necessary to check pirates and high-handed maritime competitors in time of relative peace. In war, these forces are combined with the larger ally. Without such an ally, however, they have no hope for long-term success.

3. Attack enemy commerce. Utilizing their few cruisers as commerce raiders, small navies can impress major powers with their fighting prowess, but without a major ally they cannot hope seriously to alter the outcome of an open conflict by this technique.

These strategic guidelines for the application of naval power are admittedly broad and theoretical. But they represent the questions raised by nations throughout history aspiring to use navies. By utilizing these hypotheses based on the advantage of historical hindsight, we may arrive at a basis for understanding the strategic history of navies in all times. Without such guidelines, the roles of naval forces are obscured both for layman and naval professional alike, giving naval matters an aura of mystery, which is not only unnecessary but potentially dangerous.

2

"THALASSOCRACY" AS A
HISTORICAL FORCE

"Thalassocracy" is a word introduced to Western Civilization by its earliest historians whose works survive, Herodotus and Thucydides. *Thalassa*, in Greek, means "the sea." To these men and their contemporaries, thalassocracy was a concept which meant, loosely, "maritime supremacy," i.e., the control of the sea lanes and islands by one state to insure its economic prosperity and thus its political integrity.

As the previous chapter demonstrates, I have utilized this ancient concept to test the use of the sea by all those nations which have engaged in maritime trade and naval operations. This heavily documented second chapter updates and refines the ancient idea of thalassocracy into a historical force by examining in some detail those nations which succeeded as full-fledged thalassocracies and those which fell short—and why. The very nature of each nation takes priority, while that of its government, merchant class and shipping, colonies, and navy become aspects of it.

Thalassocracy is therefore a universal concept, as indeed is its continental counterpart, the land-oriented state.

This chapter was originally presented as a scholarly paper, "The Concept of Thalassocracy in History," at Varna, Bulgaria, in 1977 before a joint meeting of the International Commissions for Maritime History and of Urban History and was published in its proceedings: *Le pouvoir central et les villes en Europe de l'Est et du Sud-Est du XVe siècle aux débuts de la révolution industrielle & Les villes portuaires* [The Central Powers and the Towns of Eastern and Southeastern

"Thalassocracy" as a Historical Force

Europe from the Fifteenth Century up to the Beginning of the Industrial Revolution, and Port Cities] (Sofia: Bulgarian Academy of Sciences, 1985). A condensed version appeared as "The Maritime World: Traders" in the summer 1987 issue of *The Wilson Quarterly*.

"Isn't it funny," American seascape artist John Marin once observed, "that Dictators *never never never* live by the sea?"[1] He was of course not raising a question but rather stating a fact: that landlubbing despots of sprawling landbound states or empires traditionally have not flourished in or near the domains of Poseidon, Greek lord of the sea, or of Portunus, Roman god of ports, and that therefore the inhabitants of port cities and maritime states are distinctive from their landward counterparts. To this concept I thoroughly subscribe.

Despite the real "danger of erecting deterministic explanations" in trying to make the past comprehensible, as Arnold J. Toynbee warned,[2] historians must nevertheless continue to attempt to classify such special physical relationships as the sea to human affairs, just as, for instance, geographer Ellsworth Huntington tried to do with the role of climate, historian Walter Prescott Webb with the frontier, and "anthropo-geographers" Friedrich Ratzel and Ellen Churchill Semple with the natural environment in general.[3]

Traditionally, the question of the historical influence of the sea has usually been the preserve of naval historians trying to underpin their strategic theories and treatises, as did Alfred Thayer Mahan in 1890 and this writer more recently.[4] But the work of many historians and ethnologists since Mahan's day, particularly over the past decade and a half, has made possible a full thalassocratic (*thalassa*, "the sea") hypothesis. Simply stated: the role of the sea in human affairs has been a special one with its own unique characteristics, and genuine thalassocracies may be found only six times in history—the ancient Minoans (c. 1600–1400 B.C.), ancient Athens (c. 500–400 B.C.), late-Medieval/early-Renaissance Venice and Florence (c. A.D. 1200–1500), the modern Netherlands (c. 1600–1680), modern Britain (c. 1650–1900), and the contemporary United States (since 1900).

As a hypothesis, such a thalassocractic factor in history raises a great many questions and concurrent demands for data to illuminate and test details, an undertaking beyond the scope of the present essay. Nevertheless, general guidelines may be established here, synthesized from the tentative findings and conclusions of several scholars, in order to understand thalassocracy, as chronographer Molly Miller has suggested, as "a concept unifying a treatment of Universal History."[5] This undertaking will also update the theories of Semple, who said in 1911, "Because adaptation to the sea has been vastly more difficult than to land, commensurate with the harder struggle it has brought greater intellectual and material rewards. . . . Hence history has always staged its most dramatic acts upon the margin of seas and oceans; here always the plot thickens and gives promise of striking development."[6] The scope is therefore broad, with environment a key element always.

As in the introductory chapter, the present hypothesis contrasts the thalassocratic state with the continental power as the more politically and religiously liberal, economically competitive and wealthy, technologically innovative and advanced, industrially sophisticated, socially cosmopolitan and diversified, intellectually creative and thus culturally dynamic. As such, the relative progressiveness of peoples with a close physical proximity to the sea has been a major element not only in the national experience of each such people but for civilization as a whole. It is not necessarily *the* sole determining force but rather *one* important catalyst for change and growth.[7]

❋ ❋ ❋

While a consideration of land-oriented or continental peoples is not within the purview of this paper, a brief characterization is necessary because of (1) the inevitable relative contrasts and (2) the fact that the great seafaring communities originated on the land. Tied to the land and a basically agricultural economy, baronial lords and their peasant tillers of the soil required centralized and authoritarian command and controls in order to insure economic, religious and political survival and stability and thus also military security. The result was fixed and expensive fortified towns and roads, standing armies and alliances, landed- and soldier/cleric-aristocracies with their despotic monarchs, frequent invasions and wars, and strict internal political regimentation and servility. To be sure, many such states had short coastlines (in contrast to their much longer land frontiers), but—in the words of Scandinavian maritime ethnologist Olof Hasslöf—enterprising fishermen and shippers on these coasts were condemned by "agrarian" and interior "town politicians" because the sea trades "took workmen away from the rural districts" and "tempted people to trade outside the towns. . . ."[8]

In landlocked states, obviously, overland communications monopolized all direct trade, but even where a coastline existed with its option of waterborne intercourse, the continental government maintained a preference for roads as long as they seemed even remotely profitable. This bias—natural as it was—reflected the continental leaders' ignorance and distrust of (and contempt for) the sea, middle class shipping merchants, navies and all the uncontrollable economic practices and social behavior of the sea-oriented citizens, not least their relatively advanced cultural diversity.[9] Viewing trade communications as "the infrastructure of all coherent history," Fernand Braudel has discovered, for example, that in the sixteenth-century Mediterranean littoral—a region that "has never been inhabited by [any] profusion of sea-going peoples"—"wheeled vehicles and pack animals" on roads successfully competed with shipborne traffic, especially between the Mediterranean and northern Europe.[10] Notes J. H. Parry of southern Europe, "Land transportation of bulky commodities, except over short distances, was prohibitively expensive," but there were many examples "of traditional overland trade routes persisting, remaining in busy use, long after more economical, or seemingly more economical sea routes had been opened. The sea routes usually won in the end, but the process of competition might extend over centuries."[11] In the meantime, as the importance of the sea grew, land-oriented states preferred to rely on foreign shipbuilders, merchant carriers and entrepreneurs for any oceanic enterprises and to thus discourage the other manifestations of thalassocracy.

The emerging seacoast city became the focus of the native seafarers' escape from such land-oriented confinements, to achieve what Frederic C. Lane has called "local independence."[12] Says Hasslöf, "Craftsmen, traders and sailors were in a better position to gain their freedom. . . . For them the [coastal] towns became a sanctuary where they could escape serfdom and other feudal agrarian restrictions."[13] For "an active coastal region" to develop, in Braudel's view, "A big city was indispensable with its supply of yards, sail-cloth, rigging, pitch, ropes, and capital; a city with its tradesmen, shipping offices, insurance agents, and all the other services an urban centre can provide. . . . The history and the civilization of the sea have been shaped by its towns."[14] The coastal cities surmounted the laws, privileges, doctrines and policies of the continental central government either by obtaining exemptions "from existing privileges, taxes and regulations concerning rural commerce and peasant shipping etc., by means of petitions and letters of complaint" or by defying said rules with great "ingenuity and subtlety" through special contracts and exceptional preferential status.[15] Then, too, such ocean traders refused to obey the mercantilistic tariffs, closed imperial barriers and wartime paper blockades of the continental rulers whose "arbitrary laws . . . could not be extended effectively to control humans beyond their shores and [be] put upon the seas." In the colorful words of human engineer R. Buckminster Fuller, such free trading, smuggling and privateering outlaws were the "Great Pirates" or "sea mastering people" of history; "the only laws that could and did rule them were the natural laws—the physical laws of [the] universe which when tempestuous were often cruelly devastating. High seas combined with nature's fog [,] and night-hidden rocks were uncompromising."[16]

The seafarer thus developed unique ethnological characteristics as an individual and in his community affairs. A rugged, "salty" and outspoken individualist faced with the natural challenges before him, he had to command a "high proficiency in dealing with celestial navigation, the storms, the sea, the men, the ship, economics, biology, geography, history, and science."[17] Consequently, such persons appear, to maritime historian G. Krogh-Jensen, as "free men, bold and wise, courageous and reliable, unimpressed and not without a sense of humor."[18] Uninterested in the politics of the landowners, these fishermen and oceanic traders focused their activities on the art and profits of seafaring. Observed the Athenian Pericles (according to Thucydides), "Seamanship . . . is not something that can be picked up and studied in one's spare time; indeed, it allows one no spare time for anything else."[19] The seafarers' accumulation of capital and material goods provided the means for economic and hence eventual political power by making possible merchant and sailor guilds which led to a new middle economic class, the bourgeoisie. Using Hanseatic societies and towns as his model, Hasslöf has shown how these men formulated "regulations that were later included in government legislation and administration. The capital association of the merchant shipping companies acquired a central and dominant position in law."[20]

Thalassocracy—the status of a full-fledged maritime state—was not achieved, however, without geographic insularity. Unless the maritime element or the state itself was actually or virtually isolated from overthrow or domination by the continental ruling element or foreign enemies—unless it enjoyed what C. Vann Woodward has termed "free security"[21]—and was situated along convenient shipping lanes for trade, it could not break away to evolve independently as an ocean-oriented state. Thus were the dynamic maritime elements of, for example, ancient Phoenicia (and Carthage), the Medieval Islamic states, the Baltic peoples of Sweden, Denmark and the Hanse, modern France and Germany kept in subordinate roles. For thalassocracy, says Miller taking her cue from Herodotus and subsequent Greek writers, "is not merely an objective fact—the most prosperous maritime state of [a given] time; thalassocracy means the possession of a fleet and an aim, a concentration of force and purpose."[22] Mahan tried to classify such a purpose in his definition of "sea power"—the sum total of production, commercial and naval shipping, and overseas markets and colonies.[23] Yet thalassocractic purpose was much more—the thrust of a vibrant, expansive, flourishing, individualistic, cosmopolitan society, set apart always by its uniquely favorable geographic position and insularity; in Semple's words, ". . . everything that enhances or diminishes the cultural possibilities of a sea—its size, zonal location, its relation to the oceans and continents— finds its expression in the life along its coasts."[24]

Whether the grand designs and exuberant energies of the thalassocracies were due to inherent psychological or acquired social-geographic-environmental ingredients, historians and other observers from earliest Greek times have discerned that such maritime states have differed from landward nations and that their influence in history has been one of growth and expansion.[25] Declares Semple, ". . .

the progress of history . . . [is] attended by an advance from smaller to larger marine areas. Every great epoch of history has had its own sea, and every succeeding epoch has enlarged its maritime field."[26] Such an expansion (or extension) Braudel—in considering the Mediterranean Sea—has articulated as a "physical law [which] seems to have operated more or less regularly."

> It is easy to imagine and even probable that the life of the sea, a vital force, would first of all have taken control of the smallest and least weighty fragments of land, the islands and coastal margins. . . . Growing more powerful and compelling, this force would draw into its orbit the larger land masses, the peninsulas, elevating the history of the sea to a higher level. And the greatest moments of all would be when it was strong enough to attract towards it the great continental blocs. . . .[27]

From the early seafaring Greeks of the Aegean down to "the final epoch of the World Ocean" in the twentieth century, mankind expanded seaward both physically and intellectually. Says Semple, "The gradual inclusion of this World Ocean in the widened scope of history has been due to the expansion of European peoples, who for the past twenty centuries, have been the most far-reaching agents in the making of universal history." Where she sees their colonization enterprises as having "something . . . cosmic" about them, Fuller labels such modern Europeans as "world men who lived on the seas."[28]

Before defining the thalassocracy by its several specific characteristics and through the examples, it is necessary in light of the above to note the general influence of the thalassocratic state both in space and in time. The *focus* of thalassocracy has naturally been understood in terms of the particular city- or nation-state(s) at the moments when its material and cultural wealth coincided, e.g., Athens during the fifth century B.C. In *space*, however, the economic, political, naval and artistic benefits of that state could very likely be extended to embrace neighboring and even distant peoples who then willingly, unwillingly or unconsciously shared the fruits of the thalassocracy. Thus the Athenian Empire policed the Aegean Sea and included allies, subject cities and colonies, while cultural Hellenism conquered as it advanced—to Macedonia, Sicily and beyond and later via Alexander as far as India and via the Romans over the entire Mediterranean basin. In *time*, consequently, the cultural vitality could survive the political, economic, and naval demise of the state itself both within that state and among the peoples who profited from their cultural contact with it directly or even indirectly. The Hellenistic world is a good example, but also even much later when the achievements of the Hellenes provided much of the stimulus for the Italian Renaissance two millenia after the Peloponnesian War had crippled the Athenian state. These same time and space factors were equally true of the other major thalassocracies.

* * *

In examining the major thalassocratic states, several ingredients or manifestations appear more or less common to all of them and at the same time distinctive from the features of agrarian continental societies. In particulars, each maritime state had many unique characteristics, but the similarities of all of them far outweigh the differences between each.

The thalassocracy's strong bourgeois middle class of merchants, entrepreneurs and capitalists prevailed over the smaller aristocracy and peasantry which however predominated in the agrarian state. This class had to be *genuinely* rooted in maritime concerns; says Mahan, ". . . aptitude for commercial pursuits must be a distinguishing feature of the nations that have at one time or another been great upon the sea."[29] Such a society was fairly egalitarian, as pictured by the sixth-century Roman Cassiodorus in his appreciation of early Venice where "rich and poor live together in equality. The same food and similar homes are shared by all; wherefore they cannot envy each other's hearths, and so they are free from the vices that rule the world. . . ."[30] Whether a budding thalassocracy or a maritime empire, one key unifying element was, in Parry's words, "the orderly regulation of trade, in accordance with fixed economic principles or assumptions [as] a major object of policy . . . serving the exclusive interest of its own subjects. . . ."[31] That economic regulation, significantly, was not the province of the central government as in continental states, but of an influential association of private business specialists, "an entrepreneurial group," argues Peter Mathias, "the shock troops of economic change . . ., the men under whose charge new sectors of the economy could be developed and innovations brought into productive use."[32] The economic form of the maritime state was therefore capitalistic—in Lane's definition "a society so organized that men can make money by investing their capital."[33]

Unlike the Marxian view of the economic motivation being primary, recent historians have assigned the "shaping and reshaping [of] business forms and relationships," in Thomas C. Cochran's words, to the factors of "social structure and cultural values" with their consequent accelerated "technological change or invention"—"continuously active forces in daily economic operations and in the longer-run processes of economic growth."[34] In their thrust to achieve independence from their feudal overlords, the maritime interests thus formed "new classes—merchant capaitalists, shopkeepers, craftsmen, and day laborers," leading to a "commercialized atmosphere," says Lane; the result was—in Renaissance Italy at least—republicanism: "the rejection of hereditary kingship in order to devise other forms of government that their creators believed would permit and encourage more men to participate more actively in making laws and choosing leaders."[35] From such favorable geographic, social and economic conditions therefore the thalassocracies took on the political character of free—or freer—systems, whether as republics like Renaissance Venice, Florence, or Holland or as democracies like ancient Athens or modern England with its American offspring.

Free enterprise and free government in the thalassocracy necessarily placed value on the individual initiative of the civil administrator-strategist-economic/political manager, who functioned in sharp contrast to the feudalistic land/warlord of the agrarian monarchy. Command and control over merchant marines and navies required not only managerial intuition and strategic expertise but sufficient decentralization in order that fleet and merchant convoy commanders could exercise flexibility in decision making across the vast expanses of water far from the geographical center of the state, waters endlessly hazardous due to physical dangers and human enemies. The same rigid controls which typified the overland trading and military routes and fixed fortifications of the continental states simply could not be applied at sea—a fact which true thalassocrats understood and their imitators did not. Maritime economic and naval power worked subtly over long periods of time, its strategic uses applied sagaciously by expert managers the likes of, for example, the Athenian Pericles, the Dutchman Jan de Witt and Englishmen Samuel Pepys and the two William Pitts. Such a landward imitator as Jean Baptiste Colbert of France served as an effective civil administrator who tried to force his country into a more maritime mold, but he could never overcome the dominating continental aspirations of his king, Louis XIV.

The freer reign of the thalassocracy's leaders was reflected and shared by its citizens but especially the subordinates and crewmen at sea, leading to a national self-confidence unparalleled in the agrarian society with its less-skilled, cheaply exploited and expendable masses of people. The sailor, merchant as well as naval, developed raw stamina and a high sense of duty—ship, service and political loyalty—a virtue which Semple elevates into a unifying national characteristic.[36] Consequently, this seaman was an outspoken individualist of the first order—rugged, practical, self-disciplined,[37] devoted to a sense of justice and fair play,[38] imaginative, tolerant, forebearing of extreme hardship and trial, and impatient with prolonged tyranny.[39] The calling of the seaman was that of a free man and citizen, honored by each of the thalassocratic governments, which in ancient and Medieval times generally refrained from employing slaves on board their warships.[40] Preoccupied with their profession and the quest for profits, the merchants, admirals and sailors of the thalassocracy alike tended to be virtually apolitical, dedicated only to the status quo which supported their activities, in contrast to the agrarian state where landward traders and army officers remained tied to fixed estates and roads, logistical bases and lines and fortified towns, a physical presence on home soil which involved them inextricably as important political elements within the civil government.

Consequently, the thalassocracies all evolved into catalysts of change—again, in relative contrast to their continental counterparts—as explorers of the physical and intellectual worlds. In the quest for new markets and virgin seas or lands, these rugged individualists, epitomized by the great mariners from the Greeks to the Renaissance, shared what Samuel Eliot Morison has termed a common "restlessness."[41] They developed maritime industries, navigational techniques and engineering sciences with which to utilize the sea and thereby stimulated a generally high level of technology and industry; they were, in short, masters of invention and resourcefulness.[42] Furthermore, since they were preoccupied neither with the fixed confines nor the defensive requirements of landward peoples, the thalassocracies ideally sought to create an ordered world not through political or military subjugation but rather through enlightened government and secular international laws of commercial intercourse.

Simultaneously worldly by their travels and religious by their exposure to the forces of nature, the maritime peoples also channeled their exuberant energies into the creation of a better life for the average citizen, and, from their own free spirit and wide contacts with foreign ideas, as opposed to the blind religious orthodoxies of the agrarian states' monarchical clericalism, they tended to be protestant and cosmopolitan.[43] Rather than advocating the blatant repression of their own poor and helpless native and colonial subjects, and indeed regarding large standing armies as superfluous for internal control or home defense, the thalassocracies tried (however imperfectly) to promote the literacy,[44] international law,[45] education[46] and general humanitarianism[47] that eventually reversed the traditionally accepted tide of slavery,[48] piracy,[49] smuggling,[50] disease, starvation and ignorance.[51] Instead of agrarian simplicity and backwardness, they fostered technological and industrial diversity and progress.[52]

Finally, perhaps the ultimate measure of thalassocratic greatness lay in artistic and intellectual creativity. Contrary to the suppression of free thought, ideas and criticism as in the feudalistic and authoritarian agrarian societies, the thalassocracies became monumental centers of aesthetic beauty and of general higher learning and cultural activity for native, immigrant and visiting thinkers and artists alike. Such superior minds exist in all societies, to be sure, but the difference of the thalassocracies lay in the bourgeoisie, the merchants and craftsmen who contributed materially to cultural growth, as during the Italian Renaissance, "craftsmen because they form the milieu from which artists often come, merchants because . . . they are often quick to patronize new arts."[53] Their motivations can be deduced from the above characteristics of any restless, cosmopolitan, life-enjoying society. The thalassocracy in decline like in later sixteenth-century Italy or the imitator like seventeenth-century France betrayed their non-thalassocratic leanings with the existence not of a real bourgeoisie but of a "pseudo-bourgeois" class which supported the arts only with a view toward being accepted into the aristocracy of a landward society.[54] Indeed, the intellectual climate of the great thalassocratic cities was so stimulating as to attract away leading minds from the less-hospitable, continental despotisms.[55]

Granted, the thalassocracies often were downright clumsy and overbearing—exploitative, brutal, shortsighted, "imperialistic" in its worst connotation, and eventually overextended and even decadent, especially when in the later histories of some, notably Athens and Venice, they shifted their major goals away from maritime to continental aspirations. But on balance it was they rather than the agrarian/continental powers that led in human progress and improvement by deed and example. Such negative characteristics as just noted were the more attributable to the landward states, which in addition to their relatively simple economic conditions and despotic political systems were more culturally sluggish because, according to Semple, they were too homogeneous, coming "into contact with people of kindred stock, living under similar conditions of climate and soil to" their own. By contrast, "The distinctive value of the sea is that it promotes many-sided relations as opposed to the one-sided relation on the land."[56] What is more,

> ... had the proportion of land and water been reversed, the world would have been poorer, deprived of all these possibilities of segregation and differentiation, of stimulus to exchange and far-reaching intercourse, and of ingenious inventions which the isolating ocean has caused. Without this ramifying barrier between the different branches of the human family, these would have resembled each other more closely, but at the cost of development. The mere multiplicity of races and sub-races has sharpened the struggle for existence and endowed the survivors with higher qualities.[57]

¤   ¤   ¤

History and the Sea

If any ideal thalassocratic state could be described, the Athenian Plato attempted it for the legendary Atlantis. Whether that state was the product of his own fertile imagination (perhaps projecting his own Athens back in time), as traditional classicists insist,[58] or actually a confused rendering of the Minoans, as many modern archeologists convincingly argue,[59] or even something else is beside the point here, for Plato did describe a thalassocracy and its most noble characteristics. Though cloaked in Greek mythological explanations—here expunged—the description resembles most of history's great thalassocracies at their peak of their grandeur:

> For many generations,. . . [the people of Atlantis] were obedient to the laws . . .; for they possessed true and in every way great spirits, uniting gentleness with great wisdom in the various chances of life, and in their intercourse with one another. They despised everything but virtue, caring little for their present state of life, and thinking lightly of gold and other property, which seemed only a burden to them; neither were they intoxicated by luxury; nor did wealth deprive them of their self-control; but they were sober, and saw clearly that all these goods are increased by virtue and friendship with one another. . . .[60]

The empire of Atlantis supposedly embraced many islands "in the open sea," had wide contacts abroad and thus received "many things . . . from foreign countries," had opulent temples of great beauty, and "the docks were full of triremes [warships] and naval stores" supporting a considerable fleet of 1200 ships; they worshiped the god Poseidon, Greek lord of the sea.[61]

The decline of Atlantis came when "human nature got the upper hand" and the Atlantids "behaved unseemingly" until they tried, unsuccessfully, to conquer the Mediterranean, land and sea alike.[62] Plato's contemporary Isocrates used almost the same words to indict his fellow Athenians for their excesses in trying to conquer both land and sea and thus to explain their decline (which may or may not link Athens to Atlantis):

> . . . they readily obtained command of the sea, whereas because of the arrogance which was bred in them by that power they speedily lost the supremacy both on land and sea. For they no longer kept the laws

which they had inherited from their ancestors nor remained faithful to the ways which they had followed in times past. . . .[63]

The glory and demise of the culturally if not politically unified Minoans may in fact be the basis for the Atlantis story, for the mid-second millenium B.C. (c. 1600–1400) maritime cities on Crete, Thera, and other Aegean islands are emerging from the archeologist's spade as the first major thalassocracy of accepted record, just as Thucydides originally claimed.[64] H. G. Wells explains their entire millenium of c. 2500–1400 B.C. as "an age of peace and prosperity unexampled in the history of the ancient world. Secure from invasion, living in a delightful climate, trading with every civilized community in the world [probably only indirectly with some], the Cretans were free to develop all the arts and amenities of life." Indeed, from the Minoans' geographical immunity from invasion and consequent artistic creations, "as admirable as any that mankind has ever produced," Wells articulates the advantages of geographic insularity, or free security: "Whenever men of almost any race have been comparatively safe in this fashion for such a length of time, they have developed much artistic beauty. Given the opportunity, all races are artistic."[65] Being the first real maritime power, the Minoan cities did not have a finished navy or formal political empire in the later and more familiar sense, but by the sixteenth century B.C. they had achieved general commercial, cultural and naval dominance over the southern Aegean and central Mediterranean, suppressing piracy and establishing trading enclaves as far away as Egypt, Ugarit, and probably Libya.[66] The Minoan decline seems to have been due to a combination of overextended naval activities and imperial aggressiveness, the rise of strong maritime competition from the mainland Mycenaeans and especially the devastation following the Krakatoa-like eruption of the volcano on Thera.[67]

The continental rivals and maritime competitor-successors of the Minoans during the last half of the second millenium B.C. could imitate and eventually fill the vacuum left by the fall of the Minoan commercial empire, but they could not begin to match what William H. McNeill has called "the lightness, exuberance, and grace of Minoan art." Contrasting the art of the Minoans with that of the continental Hittites and even the earlier Mesopotamians, he detects "a fundamental difference between a land-based military empire of the second millenium B.C. and the gayer, freer life sustained by riches raked from the sea."[68] The Mycenaeans of mainland Greece replaced the Minoans on Crete and elsewhere, parroted much of their culture and extended their trade routes even further, but the Mycenaean city-states and culture seem to have been continental and warrior-oriented, eventually (c. 1200 B.C.) succumbing to overland migrations.[69] Continental New Kingdom Egypt retreated before the Hittites, both of them relying upon the shipping of Ugarit and other port cities of Canaan, although Jack M. Sasson's use of the phrase "Canaanite Thalassocracy" is primarily economic and thus inadequate.[70] Migrating "sea peoples" conquered the Canaanite coast in the thirteenth and twelfth centuries B.C. and from whom evolved the independent port city-states of the Phoenicians. The latter, centered at Sidon and Tyre, were a dynamic commercial and colonizing people—witness Carthage—but remained subject to the continental conquerors Assyria and Persia during the first half of the first millenium B.C.[71] Their Hebrew neighbors, less receptive to political submission, maintained a landward and strong warrior base, "had very little connection with the sea, and even disliked or dreaded it."[72] Indeed, no true thalassocratic successor followed the Minoans for nearly another millenium, until the Aegean world rallied around Athens in the mid-first millenium B.C.

Athens, the prototype of all subsequent thalassocracies, has been the subject of historical debate ever since Athenian statesmen and scholars first extolled the virtues of their unparalleled culture, but few historians today would argue with Athens' role as a wellspring of Western civilization. Evolving within a general Aegean civilization dominated first in the seventh century B.C. by maritime Corinth and late in the next century by such island city-states as Samos and Aegina,[73] Athens throughout most of the fifth century controlled Aegean political relationships with her seaborne empire and fleet and served as "the cultural centre of the Greek world," a development— says Russell Meiggs—"closely linked with the growth of Athenian power and wealth."[74] Following the example of Themistocles who defeated the Persians and then created the initially voluntary Delian

League, Pericles completed the subjugation of its members who resisted Athenian imperial control and thus set the stage for 2500 years of historiographical criticism of Athenian excesses.[75]

No greater advocates and defenders of Athenian thalassocracy have existed than the Athenians themselves. Into the mouth of Pericles, Thucydides not only placed some shrewd definitions and marked advantages of sea power but also the interrelationships between Athenian strategic insularity, democracy and high creativity: "Our city is open to the world. . . . Our love of what is beautiful does not lead to extravagance; our love of the things of the mind does not make us soft. . . . Taking everything together . . ., I declare that our city is an education to Greece. . . ."[76] Like Thucydides, Isocrates defended his city's imperialism as a bullwark of democracy: "We helped the common people and were declared enemies of narrow oligarchies, for we thought it monstrous that the many should be subject to the few. . . ."[77]

Modern scholarship tends to agree in the overall positive assessment of Athens. "On the political plane," says Toynbee, "the Athenian industrial and seafaring population constituted the electorate of the Athenian democracy. . . ."[78] Upon this base arose the cosmopolitanism of this thalassocracy. In the words of McNeill:

> . . . as Athenian citizens encountered foreign places and strange customs in the courses of their ventures overseas, and as strangers crowded in upon the city which had become the mercantile mistress of the eastern Mediterranean, . . . reflective minds [were provoked] to speculation, challenging them to interpret afresh and for themselves the nature of the universe and of man. With emotions tied securely to a familiar social frame, and with minds free from commitment to any particular view, the Athenians were thus ideally situated for cultural creativity. The admiration of all subsequent ages attests the use they made of their opportunity.[79]

Key to the intensiveness of this intellectual growth, in time and space, T. B. L. Webster has shown, was a pervasive spirit of competitiveness, even among the artists in their "small city with many meeting places, an open society, a viable social structure, intelligent patronage, educated audiences, the spirit of competition. . . ."[80]

History and the Sea

No thalassocracy competed with or succeeded Athens in antiquity, which helps to explain the lack of political stability and sustained cultural vitality in the middle sea after Athens' demise at the end of the fifth century B.C. Thucydides' indicting contrast placing maritime Athens over continental Sparta may be applied over the continental Hellenistic and Roman states as well. The vehicle for his statement was the neutral but once thalassocratic Corinth, whose spokesmen were addressing the oligarchical Spartans:

> An Athenian is always an innovator, quick to form a resolution and quick at carrying it out. You, on the other hand, are good at keeping things as they are; you never originate an idea, and your action tends to stop short of its aim. Then again, Athenian daring will outrun its own resources; they will take risks against their better judgement, and still, in the midst of danger, remain confident. But your nature is always . . . to mistrust your own judgement, however sound it may be . . . [and] while you are hanging back, they are always abroad; . . . they prefer hardship and activity to peace and quiet. In a word, they are by nature incapable of either living a quiet life themselves or of allowing anyone else to do so.[81]

The intellectual manifestations of this restlessness, having fashioned the golden age of Pericles, went on to conquer Alexander the Great and his Successor states and finally Rome, but their societies were continental and despotic and could only imitate that high Hellenic culture and its maritime aspects. First Sparta, unthalassocratic and unchanging in spite of its victory in the Peloponnesian War, relinquished maritime supremacy again to Athens, which then further dissipated its thalassocratic energies in more imperialistic wars. The same may be said of Hellenistic Alexander and his *diadochi*; notes M. Cary, "Between the Athenian thalassocracy of the fourth century, and the Rhodian thalassocracy of the second, none of the Hellenistic powers took adequate steps to police the seas. . . ."[82] Hellenistic art "was essentially secular,"[83] unlike that of Athens; Alexander, moving eastward, created "Greek communities, where Greek law, Greek worship and Greek habits of life prevailed,"[84] and the Ptolemies of Egypt made Alexandria "into a rival of Athens for the intellectual primacy of the Greek world," a city which "with its cosmopolitan population, and its geographical position, was a clearing house of ideas, as well as of trade, and . . . sensitive to the movements of world thought. . . ."[85] But these continental successors of Athens could not begin to originate a thalassocracy in cultural or maritime dimensions even approaching those of their teacher.

Neither could Rome. Initially sharing the Western Mediterranean with the Phoenician Carthaginians and Italian Etruscans (and Greek colonies), both essentially continental and warlike peoples[86] with initially considerable fleets by comparison, the landed-warrior Romans eventually developed, in Lionel Casson's words, into "an anomaly in maritime history, a race of lubbers who became lords of the sea in spite of themselves."[87] No analysis of Roman "land-lubberism" has surpassed that of J. H. Thiel, whose essay "The Romans and the Sea" is a model of thalassocratic historiographical analysis.[88] He views Roman recklessness toward and on the sea as clear evidence of the Roman land-lubber: "The Romans are not afraid enough of the sea, because they do not know and therefore underrate its dangers." Judging Roman nautical poetry as that of the bather rather than of the seaman, Thiel regards the erratic naval successes of the Roman Republic as the result of a sagacious use of Italiot and other allies, army-style legionary boarding tactics and the Romans' "imperturbable morale," "indomitable energy and supple litheness. . . ." Roman citizenship was above service on war or merchant vessels; the fleet served essentially to transport the army; and the aristocracy had a typical anti-maritime bias, for "Rome always was at heart a typically continental and not a naval power and so she stuck to the *minimum* in the naval sphere." And, again: "the Roman mentality always remained mainly continental, *not* thalattocentric." And since naval demobilization inevitably followed victory in war, Thiel concludes, ". . . is it not a typical symptom of a non-thalattocentric mentality and of a complete absence of attachment to the sea, if a nation, which pretends to rule and keep together a world empire round the Mediterranean, plainly withdraws from that sea, as soon as it has no longer direct and manifest naval adversaries within its horizon?"[89]

Even though "a standing, carefully and continually kept-up navy was altogether out of the question before the reign of Augustus,"[90] improvised Roman fleets during the third century B.C.

helped importantly in the conquest of the Western Mediterranean, during which however the Roman army "never got its sea-legs. To the Romans the sea was something incalculable, treacherous."[91] The overthrow of Alexander's successors in the East was no different, and even the very useful local and cooperative thalassocracy of Rhodes in the second century B.C. was soon overturned by the power-hungry and impatient Romans. Throughout the Republic, F. E. Adcock observes, there was simply "no active Roman thalassocracy and no continued effort to police the seas."[92] By the time of Caesar Augustus, the middle sea had become a Roman lake, policed by an army-dominated anti-pirate navy that acted also as a political counterweight to the legions for several subsequent emperors.[93] The maritime tranquility of the first two centuries of the Empire contributed to Rome's cultural golden age and encouraged overseas trade, much of it Greek operated and largely the result of Augustus' enlightened middle class origins.[94] But the Roman naval establishment remained as a deterrent force, for, according to the later Empire observer Vegetius, "none will attack or insult a [naval] Power, known to be always ready, and prepared to avenge any Affront."[95] Hellenized Rome never aspired to, nor did it attain, thalassocratic greatness.

From the thalassocratic frame of reference, the Middle or Dark Ages are aptly named, for though many states West and East enjoyed maritime success and even reflected related cultural influences from seaborne enterprise, none of them approached the preceding Athenian Hellenism or the subsequent Italian Renaissance in thalassocratic grandeur. Instead, continental states—land-oriented, aristocratic, authoritarian, religiously orthodox and army-dominated—prevailed everywhere as feudalism replaced the city-states of antiquity. In Europe, inward-looking but militant Christianity accompanied the Roman demise but also provided the great strength of the Byzantine Empire, "one of the most powerful conceptions of monarchical authority ever known."[96] Byzantine maritime and naval activities against the Arabs before A.D. 1000 no doubt influenced the golden age of Byzantine culture, but the Empire remained ever authoritarian, especially after 1000; "the urban element underwent a comparative retrogression with the decay of naval power, while the dispersal of land forces throughout the countryside in effect magnified the power of the provincial aristocracy."[97] Christianity also absorbed and thus stifled the wild overseas plundering of the Vikings, a combination which however helped to sow the seeds of the eventual general European Renaissance.[98]

Outside the Christian world, thalassocracy never matured, then or subsequently, save for one local exception. In the Middle East, Arab and Turkish Moslems did however utilize maritime and naval forces to expand trade across the Indian Ocean, to fight the Christians and to augment continental frontiers.[99] In the Far East, maritime enterprise affected the political, economic and cultural growth of Buddhist Gupta India (fourth to sixth centuries)[100] and late Sung, Yüan and early Ming China, but not sufficiently to overcome their traditional continental biases.[101] The exception was the Indonesian empire of Srivijaya (seventh to fourteenth centuries); argues G. W. Spencer, it was "a true thalassocracy, oriented almost exclusively toward the sea." As a "loosely-integrated network of trade-entrepots" focused on Palembang, Srivijaya dominated the Straits of Malacca, "the great maritime trade funnel of Southeast Asia," and evolved into "a cosmopolitan, urbanized, culturally-receptive trading community" heavily influenced by Indian Buddhism and culture in general.[102] Overall, however, Medieval continental peoples and their religions throughout the world eventually stifled any promise of continuous thalassocracy into the modern period.

❉   ❉   ❉

Thus when the Renaissance began, it was ushered in not by such feudalistic continental states but by new thalassocracies pointing the way into the modern era. Herbert Rosinski, for example, discerns a critical factor in the creation of the modern nation state as "the opposition between continental states on one side and maritime states on the other, or between absolute and constitutional states." The latter placed the individual ahead of the state, so that "the establishment of the constitutional state led logically and directly to modern democracy."[103]

The Renaissance commenced in Italy among the several republican city-states, the leading ones being commercially and maritime oriented.[104] Such cities not actually located on the coast, notably

Florence, found indirect access through their coastal neighbors, in particular Venice, Genoa, and Pisa. From about 1100 but certainly 1200, these four cities and many less-important ones closely resembled the Aegean world of the fifth century B.C. in that they constantly warred while sharing an immensely prosperous economic wealth and cultural dynamism. In addition to the commercial stimuli natural to thalassocratic impulses, they enjoyed the added benefit of the rediscovered ancient states and examples, so that their thalassocracy—always centered on Venice and Florence—combined both the old and the new to bear a striking resemblance to the Hellenism of ancient Athens. As Peter Burke puts it:

> . . . the comparison with ancient Greece, and more particularly Athens in about the fifth century B.C., has proved irresistible; for not one but a whole cluster of cultural and social traits resemble Renaissance Italy. There was an emergence of artists as individuals with distinct styles. . . . The competitive spirit was strong. Cities were large, governed themselves, and patronized the arts. In Greece as in Italy, there was a linguistic unity but not a political one; hence culture had an important unifying function.[105]

Or, in the view of McNeill, "Without the economic underpinning and the artistic and intellectual stimuli arising from this trade, the flowering of city-state culture in Italy which we call the Renaissance could not have occurred."[106]

The fulcrum of the struggles between the several Italian states and the more general conflict between the Christian Western Mediterranean and the Moslem Eastern Mediterranean became Venice, by 1300 the leading thalassocracy of the Mediterranean world, "a symbol," says Lane, "of beauty, of wise government, and of communally controlled capitalism . . ., the freest of Italy's many free cities. . . ." In contrast to the embattled Crusaders, Medieval and Moslem states of the large land masses, "It had no city walls but a lagoon, no palace guard except workers from its chief shipyard, no parade ground for military drill and display except the sea. The advantages of its site fostered also an economy which combined liberty and regulation in ways as unique as Venice's urban arteries and architecture."[107] It was "almost wholly mercantile" and "essentially a commercial enterprise," concludes D. S. Chambers,[108] and by the fifteenth century Venice was also, notes Oliver Logan, "the great European centre for hellenistic studies and for the editing and printing of classical texts in general," whose culture was "so strikingly individual . . ." that it "derived part of its character from general European and Mediterranean influences and from the tensions that her international position imposed."[109]

The fifteenth century rightly deserves its place as a watershed era, for while the cultural splendor of the Italian Renaissance did not wane, the economic and naval might began to diminish, and the continental states of Spain, Portugal, and Ottoman Turkey took to the sea though always relying heavily on the Italian merchant carriers.[110] Granted, Florence finally conquered Pisa in 1406 and thereafter went to sea on her own, leading to "an opulent and higher bourgeoisie, . . . disdaining none of the necessary tasks of commerce, industry and banking, and [which] also paid homage to the refinements of luxury, intelligence and art."[111] The social change that attended such grandeur was hardly due to the coincidence that Hermann Hefele and Braudel assign it,[112] but was rather a natural result of thalassocracy. Still, both Florentine and Venetian commercial and naval power were on the wane,[113] while both cities along with Genoa were becoming more involved with the political rise of Spain and France, especially after the latter invaded Italy in 1494. The shift of the bourgeoisie from sea to landward interests is part of the explanation,[114] but so is the desire of Spain to hire Italian bankers and seafarers to enhance its economic fortunes.[115] These trends continued throughout the following century (the Dutch replacing the Genoese financiers in Spain), with only a less-pronounced diminishing of general Italian culture.[116]

The new continental powers, notably Spain and Portugal, beset by external and internal concerns—frontier defenses, religious struggles and closed feudalistic political and economic systems—entered the modern age during these years but still clinging to Medieval ways that were barely affected by their considerable oceanic achievements. Neither Spain nor Portugal developed into natural thalassocracies, for neither had a native bourgeoisie, and Spain depended largely on foreign merchants and mariners, the latter "in a sense, international

mercenaries, *condottieri* of the sea."[117] Mahan characterizes the
Spanish government as "in many ways such as to cramp and blight a
free and healthy development of private enterprise; but the character
of a great people breaks through or shapes the character of its
government, and it can hardly be doubted that had the bent of the
people been toward trade, the action of government would have been
drawn into the same current." The fortunes of Portugal "followed the
same downward path; although foremost in the beginning of the race
for development at sea, she fell utterly behind."[118] Indeed, Portugal's
pioneering efforts at sea were almost accidental, the result of coinci-
dences of accidental historical timing and a favorable geographical
location "on the street corner of Europe." For the Portuguese, says
Parry, "were only marginally a maritime people," plagued by manpower shortages because of a small urban population, an elite landed
aristocracy and a vast majority of peasants. The country therefore
suffered from a backward economy and society and the constant
threat of attack from neighboring Castile. And yet, as soon as that
"menace was lifted temporarily from their backs, the Portuguese . . .
embarked on a career of maritime expansion . . ., very quickly discovered a taste and an aptitude for it, and pursued it with a reckless
persistence. . . ." Portugal's unnatural maritime endeavor did not
survive the sixteenth century for the internal reasons noted above, the
Spanish conquest in 1580 (until 1640) and new genuine maritime
competitors. Still, in all, the achievement was impressive.[119]

Thus despite Spain's discovery of the New World and Portugal's
rediscovery of the Orient, one may agree with J. H. Elliott that these
revelations constituted "a challenge to the whole body of traditional
assumptions, beliefs and attitudes."[120] The entrenched social and
religious customs of Iberian society successfully resisted the intellectual opportunities thus presented, leading Parry to remark, "Discovery is not necessarily a social catalyst, nor the possession of empire a
harbinger of change."[121] Consequently, the Spanish and Portuguese
of the sixteenth and early seventeenth centuries crushed the peoples
they encountered or barring that capability forced them back into
relative seclusion as in Ottoman Turkey, China, and Japan.[122]

Only by defining thalassocracy as in the present essay can one
avoid the problems encountered by Braudel in his masterful depiction

of the Mediterranean world between 1550 and 1600. These were years when no one thalassocracy prevailed or existed, a period that was indeed bracketed by the Venetian-Florentine and Dutch eras. He observes quite correctly that the peoples of the Mediterranean in that period were primarily agrarian and even migratory along the coasts, with inadequate numbers of them available to follow the occupations of the sea, with a bourgeoisie either nonexistent as in Spain or foreign as in Turkey or "on the verge of disappearing" as in Italy; such societies were therefore typified by a growing internal polarization between aristocrats and peasants.[123]

> . . . the Mediterranean in the sixteenth century was overwhelmingly a world of peasants, of tenant farmers and landowners; crops and harvest were the vital matters of this world and anything else was superstructure, the result of accumulation and of unnatural diversion towards the towns. Peasants and crops, in other words food supplies and size of the population, silently determined the destiny of the age. In both the long and the short term, agricultural life was all-important.[124]

Braudel is therefore in error by regarding "the great maritime empires" as being centered in "Istanbul, Rome or Madrid."[125] The Ottoman, Roman, and Spanish empires were simply never thalassocratic, either in purpose or achievement. There were in fact no genuine maritime states dominating the seas during the course of the sixteenth century.

The change came in the seventeenth century, say by 1650—the date at which Braudel and Elliott agree that on the one hand the Mediterranean decline became pronounced and on the other the new overseas discoveries began to affect the intellectual climate of Europe directly.[126] Though the United Provinces of the Protestant Netherlands led the assault on the Spanish and Catholic imperial monopoly, the attack was a shared one centered in the northwestern European littoral with its greater resources of nautical manpower, timber and naval stores than the Mediterranean and its more receptive attitude toward Renaissance thinking, which came thence directly from Italy during the seventeenth century. New thalassocracies resulted, the direct consequence of these intellectual challenges and those especially stimulated by the New World which, in the words of Elliott, "gave the Europeans more room for maneuver. Above all, it promoted movement—movement of wealth, movement of people, movement of ideas. . . . The opening of a new frontier on the far shores of the Atlantic therefore created new opportunities, and a climate of thought which encouraged confidence in the possibilities of success."[127]

By 1650, the Netherlands and England had become thalassocracies, while France and to a lesser extent Denmark and Sweden were trying, but in vain, to become serious maritime competitors. The failure of the latter three lay simply in the continental nature of their growing authoritarian governments and orthodox clergies; agrarian, non-mercantile class structures; exposed strategic frontiers, related standing armies and recurrent wars. As successors of the Italian Renaissance, Holland and England launched a Protestant attack on continental Catholicism, supported their own growing merchant-financier middle classes which took over the carrying trade of Europe, created great commercial-naval fleets and colonial empires, and developed parliamentary political institutions.[128] In contrast to their continental rivals, England and Holland were thalassocratic to the core; indeed, the fact that their periods of greatness overlapped during one generation, c. 1650-1680, helps to account for the viciousness of the three Anglo-Dutch wars of these years.

While internal difficulties delayed England's maritime expansion after the repulse of the Spanish Armada (1588), the Dutch lost no time in asserting their seaborne and cultural supremacy throughout most of the seventeenth century. Building on a firm Medieval economic base of independent action, the Netherlands by their proximity to the sea (and a river-canal system) were "destined to become and remain a country of sailors, fishermen, traders, and farmers," leading to, according to J. H. Huizinga, an urban middle-class, thoroughly bourgeois and democratic society, for "life at sea had a leveling effect on Dutch society as a whole." Closely related to these elements was the Dutch rejection of orthodox Christianity; "the influence of Calvin's doctrine and spirit was undoubtedly decisive in the rise and preservation of the new, free state of the United Provinces, and in the foundation of the Dutch Empire overseas."[129] Conquering the East

Indies and warring along the Spanish Main and in Portuguese Brazil, the Dutch imperial effort augmented their monopoly over the European carrying trade in the early seventeenth century. As "a pivot of internationalism—in its alliances and conflicts, its economy and culture,"[130] the Netherlands developed a "cosmopolitan outlook"[131] that embraced and encouraged Renaissance philosophy, science, the arts and humanitarianism on a vast scale, as well as the beginnings of modern international law; J. H. Plumb observes that "science and philosophy were both stimulated by the great extension of human knowledge that the geographies brought about. The trans-ocean trade brought more than a profit; it made windows into the mind."[132]

The political, economic and cultural decline of this dynamic thalassocracy, rather abrupt from the late seventeenth century, may be attributed largely to a combination of continental influences and maritime rivals. The latter came in the form of English and French attacks on Holland's seaborne vitals—by mercantile and industrial competition and naval offensives against the shipping and the overextended colonial trade routes.[133] The continental attack came from neighboring France, whose army invaded the Netherlands during the Third Anglo-Dutch War and whose own emerging and brilliant culture began to submerge and affect that of the Dutch.[134] Competition gave way to "a love of tranquility," merchants became magistrates and entrepreneurs investors.[135] Or, as Charles Wilson puts it, "economic stagnation, . . . a stagnation of spirit, a sapping of creative power, an end of greatness and a slide into social artificiality" can be explained by the physical and spiritual exhaustion of a people who having heaved "themselves up by their own bootstraps with unbelievable determination" now accepted the fact "that they were too small for their own effort to count for much in the European struggle for power."[136]

The considerable French cultural and maritime contribution to the Dutch demise is ironic since France's bid for thalassocracy failed before ever having attained the heights of Holland's achievement. Several historians over the past century have made much of France's achievements upon the sea, for example Gerald S. Graham who accords France "a great maritime inheritance," for being "in many respects an amphibious state" with "a race of superb seamen" on two coastlines that "inevitably stimulated the maritime and commercial ardour of French governments."[137] Indeed, France did create a formidable mercantile presence, an overseas system of colonies (though of dubious economic or strategic value), a respectable merchant marine and a strong fleet—and all in the midst of the cultural grandeur of the Sun King of Versailles, Louis XIV. But these accomplishments were largely the work of his enterprising Minister of State and Secretary of State for the Navy and Colonies, Colbert. A shrewd if unpopular mercantile and imperial administrator of middle class origins, Colbert actively sought thalassocracy by patronizing maritime endeavors as well as the arts from the mid-1660s till his death in the mid-1680s, at which time his son the Marquis de Seignelay tried to perpetuate his programs. If Colbert had any model, it was the Netherlands,[138] but most historians from Voltaire to Mahan to, most recently, D. G. Pilgrim, have concluded that the French effort at sea power was artificial; none have improved on Voltaire's statement that the great changes in France could not have occurred without Louis' (and thus Colbert's) catalytic force: "there would have been no fleets, no encouragement accorded to the arts . . . had there not been a ruler to conceive of such great schemes, and with a will strong enough to carry them out."[139]

The French attempt at thalassocracy was doomed from the outset by several factors: a continental defensive preoccupation (generally absent during the peak decade of the 1680s),[140] feudalistic peasant resistance to maritime service, an aristocracy antagonistic to commercial pursuits and indifferent to the peasant class, and consequently unresolvable logistical and administrative problems that frustrated the creation of a true maritime state.[141] "Such growth" at sea, concludes Mahan, ". . . was forced, and depended upon the endurance of absolute power which watched over it."[142] Equally forced was Colbert's patronizing of the arts by the central government, which though it generated much artistic creativity also constrained the kind of free literary or artistic tradition enjoyed by England on the one hand or Holland on the other. Says David Maland, Colbert's search for order "led to the impoverishment of inspiration in the arts by the imposition of too great a degree of control and uniformity," and at that forced into only one city, Paris, where everything could serve to emulate the image of the Sun King.[143] Furthermore, the bourgeoisie patronized the arts not for genuine cultural ends but only to ape the landed nobility, which betrayed a serious "lack of self-confidence"; "the bourgeoisie despised its own image. . . ."[144] Following the death of Colbert in 1683 and French naval reverses in 1692, Louis XIV lost interest in the sea, and the continent reabsorbed French energies as before. The most that can therefore be said for France's brief flirtation with thalassocracy is that it improved the position of the theretofore repressed merchant classes, not only in France but in such a "new" continental state as Prussia[145] and that it probably helped to stimulate the high French culture of the late seventeenth century. But it also left an unfortunate legacy in France of alleged naval prowess that led—throughout the eighteenth, nineteenth and early twentieth centuries—to several attempts to emulate it, all failures and essentially for the same continental reasons.[146] France was never thalassocratic.

Where Holland succeeded only temporarily and France not at all, England became a true thalassocracy in the seventeenth century and flourished as such well into the twentieth century—the longest in history. Gradually throwing off its continental aspirations as a consequence of the Hundred Years War, England under Henry VIII broke its orthodox religious fetters and began to lash out at the Iberian maritime economic monopoly during the sixteenth century, an inter-relationship seen in the person of Sir Francis Drake. "From boyhood," notes Louis B. Wright, Drake "had experienced the combined influences of Protestant religion and the sea," shown in "a hatred of the great Catholic sea power. . . ."[147] As an island nation, Drake's England enjoyed that natural geographic isolation which "encouraged the very confidence which facilitated the expansion of both industry and commerce, a confidence largely denied . . . to the other leading Western nations."[148] The attack on the Spanish colonies and later on the Dutch European carrying monopoly combined with English exploration and colonization in the New World, leading, in the view of J. Holland Rose, to "intellectual, social and political results" which were "incalculable."[149]

Mahan admires the consequent creation of the global British Empire and naval supremacy that reached full fruition during the course of the eighteenth century; the Britishers' "many wants, combined with their restless activity and other conditions that favored maritime expansion, led her people abroad. . . . Their needs and genius made them merchants and colonists, then manufacturers and producers; and between products and colonies shipping is the inevitable link. So their sea power grew."[150] Sagacious use of this power and geographic insularity preserved the British Isles from the ravages of the French wars between 1756 and 1815 and enabled British society, with its texture of "a pervasive looseness," to initiate the Industrial Revolution during the 1780s. McNeill assigns British industrial prosperity to a combination of factors: the "Protestant ethic," a "nonconformist conscience," the "capitalist spirit," convenient sources of coal and iron ore, a ready labor force, "and a class of innovators and entrepreneurs willing to develop new ideas and able to acquire the money or credit needed to employ them in new machines and procedures." The lack of more centralized control seems to have also stimulated British industrial growth.[151] Cultural creativity of the Enlightenment focused on Britain throughout the eighteenth and early nineteenth centuries, no doubt at least partly the direct consequence of thalassocracy.[152]

*   *   *

The Industrial Revolution of the nineteenth and early twentieth centuries appeared to enhance the capability of continental states to share the benefits of British prosperity, though in fact the material and intellectual freedoms enjoyed by the Christian West were in large measure due to the British command of the seas which enforced general peace throughout the era of the *Pax Britannica*.[153] The most direct benefactors of such a global stability were the United States and Germany. Both nations could industrialize, expand overland and somewhat overseas, unify in spite of immense internal political struggles, and even resist continental preoccupations to fashion strong democratic-mercantile institutions largely at the sufferance of the Royal Navy.[154] Germany, however, even with its brilliant intellectual achievements, could not overcome its "relatively rigid class system" and "powerful and popular mystique of state, army, and bureaucracy"[155] typical of continental handicaps—and was plunged into two devastating world wars under the Wilhelmian and Hitlerite despotisms, both of which misunderstood and misapplied the larger uses of sea power.[156] No thalassocracy, Germany inhibited or expelled her intellectuals and artists.

The United States, a genuine extension of British thalassocracy from the early colonial settlements of the seventeenth century, must be considered thalassocratic at least as long as "New" England continued (with agrarian Virginia) to dominate national life—politically, economically and culturally.[157] Comparing the early American republic with early Venice, Lane notes that "the sea was a source of wealth, contributing to the expansion of the rest of the economy,"[158] certainly from the Revolution until the 1830s, the period between the Presidencies of the Massachusetts thalassocrats, John Adams (1797-1801) and John Quincy Adams (1825-1829).[159] Northeastern American thalassocracy was made prosperous by a seaport-centered "business revolution"[160] and was reflected in the rich New England literary and Transcendentalist movements of the early nineteenth century.[161] The commercial character of the region, which after 1815 also included the port of New York, was essentially thalassocratic in the best English tradition, having, says Benjamin W. Labaree, a "civilizing effect, for the life of the Yankee mariner became a part of the wider world around him—the world of the Carolinas and Cadiz, of the Chesapeake, the Caribbean, and London itself."[162] When the country turned inwards from the 1830s toward continental expansion and internal industrialization, New England gradually lost its preeminence; American thalassocracy became passive, enjoying its "free security" thanks to a cooperative British navy until it developed, unilaterally, its own maritime stance after 1898.[163] Inasmuch as America then assumed the political, economic and naval stance of a traditional thalassocracy, developed its own literary Renaissance during the 1920s and thereafter became a refuge for intellectuals escaping the continental dictatorships, one may possibly rank the United States as a thalassocracy also for the early twentieth century.[164]

Beyond such a tentative conclusion, thalassocracy as a twentieth-century phenomenon is too colored with contemporary strategic and political biases and conflicts to allow fair historical comparisons with the seaborne states of old. The two superpowers of midcentury have tried to fill the oceanic vacuum left by the demise of the British Empire, but this does not mean necessarily that the Soviet Union and the United States have been evolving inexorably into maritime states. Robert J. Kerner's thesis of a determined Russian "urge to the sea" is no more than a statement of Russia's internal overland and riverborne territorial expansion *toward* physical continental limits on the Eurasian coasts, not unlike America's very similar nineteenth-century expansion to the North American coasts, although the United States had a substantial merchant marine to project its commerce overseas, and Russia played at navalism between the time of Peter the Great and the 1960s.[165] Indeed, the only current meaningful approach to assessing American and Russian thalassocratic possibilities, in lieu of adequate historical perspective, is through strategic and economic analyses.[166] But perhaps with the end of traditional seaborne expansion and its oceanic frontiers, the new realization for common cooperation, protection and policing of the oceans and the collapse of the mercantile seaborne empires, thalassocracy may in fact be a philosophical dead end for historians examining contemporary Russia and America.

Thalassocratic historians, not coincidentally, have been universal historians, generally (if guardedly) optimistic about the future of the race as well as in judging the benefits bestowed by past sea-oriented peoples upon Western and thus all civilization. While recognizing certain evils which do exist within this civilization, these historians reject the pessimism of several continental philosopher-historians to discern progress through it all. McNeill for example speaks of the *rise* of the West into an ultimate global cosmopolitanism, and like Wells and Toynbee he refutes the gloom of Spengler's and Nietzsche's *decline* of the West.[167] Whether such optimism is naive wishful thinking, the product of sound historical analysis, a combination of both and/or the harbinger of a genuinely new and different era of history and historiography remains to be seen.

Not surprisingly, however, the universal historian now finds himself in the company of equally optimistic scientists and visionaries who like Fuller view the planet earth as one "synergistic ecosphere"[168] and who see in the continued conquest of the physical universe—air and space—in the twentieth century a fresh thalassocratic impulse on a global scale and in much the same phraseology. They have taken their cue from President John F. Kennedy's original call for the exploration of "this new ocean," from which the "growth of our science and education will be enriched by new knowledge of our universe and environment, by new techniques of learning and mapping and observation, by new tools and computers for industry, medicine, and the home as well as the school."[169]

A common theme of this new genre of "space age" history is "the almost exact parallel between the expansion of Europe during the Great Age of Discovery in the fifteenth century and the exploration and expansion into space of the present day," and thus the possible repeat of, according to James G. Allen, the "rise of humanism, the beginnings of a scientific spirit, and . . . more precise instruments . . . [for] a more positive, aggressive and adventurous attitude."[170] Science writer Arthur C. Clarke, perhaps the most articulate spokesman for such future-history, envisions "new frontiers of the human mind" emerging from the conquest of the physical space frontier.[171] "Human thought will undergo another change of scale, comparable to that which occurred during the first age of Discovery, half a millenium ago."[172] Only time will tell of course whether such proclamations are mere scientific propaganda, a passing fancy or indeed part of the continuum of thalassocratic thought.

⚬   ⚬   ⚬

That historians are the product of their own historical environment is a truism of the profession. This being so, the call of the universal historian must not be regarded with cynicism and disdain, for his intuition is no less historical or human than that of his more specialized colleagues. Toynbee's arguments in fact have more meaning now than ever before: "In a world that has been unified in both space and time, a study of human affairs must be comprehensive if it is to be effective. It must include, not only the whole of the living generation, but also the whole of the living generation's past."[173] In this spirit, the present essay was undertaken in order to discern how, if at all, man's relationship to the sea has been a vital force in that past. The writer hopes that this thalassocratic hypothesis will not only generate criticism of the particulars of an admittedly broad and general argument but also that it will contribute meaningfully to the dialogue over the interrelationships between physical environment and human affairs throughout history.

NOTES

1. John Marin to Alfred Stieglitz, Sept. 1, 1938, quoted in John Marin (Cleve Gray, ed.), *John Marin* (New York, [1970]), 62. (Italics and capitalization in the original). "Don't lo[c]k the tyrants up—make them look at this . . . great-sea manifestation [a storm] taking place. . . ." Marin to Stieglitz, undated letter, 61.
2. Arnold J. Toynbee, *A Study of History*, rev. abr. ed. (New York, 1972), 477.
3. Ellsworth Huntington, "Changes of Climate and History," *The American Historical Review*, XVIII, no. 2 (January 1913), 215, stresses that climate is only "*one* of the important factors" in history (my italics). Walter Prescott Webb, *The Great Frontier* (Austin, 1952, 1964), 1, regards the Frontier "as *a* determining factor in modern Western civilization" (my italics). See, for example, Ellen Churchill Semple's elaboration of Friedrich Ratzel's theories, *Influences of Geographic Environment: On the Basis of Ratzel's System of Anthropo-Geography* (New York, 1911), arguing for more than *one* influence.
4. Alfred Thayer Mahan, *The Influence of Sea Power upon History, 1660–1783* (Boston, 1890), 25ff.; my "The Thalassocratic Determinism of Captain Mahan," in Reynolds and William J. McAndrew, eds., *Proceedings of the 1971 University of Maine Seminar in Maritime and Regional Studies* (Orono, Me., 1972), 77–85; and my *Command of the Sea: The History and Strategy of Maritime Empires* (New York, 1974), 2ff. and passim.
5. Molly Miller, *The Thalassocracies: Studies in Chronography II* (Albany, 1971), 2. Synonyms are thalassic, thalattocentric and thalassologistic.
6. Semple, *Geographic Environment*, 294.
7. On the dangers of thalassocratic determinism, an Israeli historian of ancient seafarers has warned: "The doctrine preached by this school, which one might call thalassologists, or people who place stress on the significance of the sea and the part it has played in the crystallization of human history is, without doubt, too extreme." Then, he adds, "Nevertheless, we should recognize that this doctrine rests on pretty solid foundations." Zvi Herman (tr. by Len Ortzen), *Peoples, Seas and Ships* (London, 1964), xii–xiii.
8. Olof Hasslöf, "Maritime Commerce as a Socio-Historical Phenomenon," in Hasslöf et al., eds., *Ships and Shipyards; Sailors and Fishermen: Introduction to Maritime Ethnology* (Copenhagen, 1972), 81, citing the Scandinavian experiences of the sixteenth to nineteenth centuries. Also, 73, ". . . maritime commercial and social life in Europe was for a long time integrated with a structure of society dominated by class privileges, constraints on trading and the doctrines with which these phenomena were associated."
9. Regarding mountain-bound communities, Carl Jung observes, "Mountains tend to restrict the horizons of the mind." In conversation with Robert Kroon, quoted in *Time* (Oct. 28, 1974), 30.

10. Fernand Braudel (tr. by Siân Reynolds), *The Mediterranean and the Mediter-ranean World in the Age of Philip II*, 2 vols., 2nd rev. ed. (London, 1972, 1973), I, 282, 138, 289–290.

11. J. H. Parry, *The Discovery of the Sea* (New York, 1974), 70, 124.

12. Frederic C. Lane, "At the Roots of Republicanism," in *Venice and History: The Collected Papers of Frederic C. Lane* (Baltimore, 1966), 522.

13. Hasslöf, "Maritime Commerce," 75.

14. Braudel, *Mediterranean*, I, 145, 278. To this process of urbanization, Parry adds a "more varied and lucrative employment, more interesting social life, greater economic and social freedom, more effective concentration of political power. . . ." *Discovery of the Sea*, 70.

15. Hasslöf, "Maritime Commerce," 93-96, for such practices in Scandinavian towns, and for England, Ralph Davis, *The Rise of the English Shipping Industry* (London, 1962).

16. R. Buckminster Fuller, *Operating Manual for Spaceship Earth* (New York, 1970), 21.

17. Ibid., 22.

18. G. Krogh-Jensen, *Den Thylandske skudefart* [The Thyland Vessel Traffic] (Thisted, 1967), quoted in Hasslöf, "Maritime Commerce," 94.

19. Thucydides (tr. by Rex Warner), *The History of the Peloponnesian War* (Baltimore, 1954), 94.

20. Hasslöf, "Maritime Commerce," 109.

21. This appropriate term, refering to the United States up to 1945, is taken from C. Vann Woodward, "The Age of Reinterpretation," *The American Historical Review*, LXVI, no. 1 (Oct. 1960), 1–19. See also Mahan, *Influence of Sea Power*, 29ff.

22. Miller, *The Thalassocracies*, 45. Her subject is the maritime states of the Aegean Sea during the late sixth and early fifth centuries B.C.

23. Mahan, *Influence of Sea Power*, 28, 71 and passim.

24. Semple, *Geographic Environment*, 282.

25. On instinct versus environment, see Toynbee, *A Study of History*, 96: ". . . the decisive factor . . . is the spirit in which Man responds to the challenge of the sum total of Nature that has become Man's environment as a result of Man's own appearance in the Universe." He sees, 95, the ancient Greeks explaining "the manifest differences between themselves and their newly discovered neighbors as being the effects of diverse environments upon a uniform human nature, instead of seeing in them the outward manifestations of a diversity that was somehow intrinsic in human nature itself." Semple, *Geographic Environment*, 22: "Environment influences the higher, mental life of a people chiefly through the medium of their economic and social life; hence its ultimate effects should be traced through the latter back to the underlying cause." J. R. L. Anderson has sought to understand one key "instinct", that of exploring, "some factor in man, some form of special adaptation, which prompts a few individuals to exploits which . . . are of value to the survival of the race . . .," but he makes no distinction between landward and oceanic explorers. *The Ulysses Factor: The Exploring Instinct in Man* (New York, 1970), 17, 23–31.

26. Semple, *Geographic Environment*, 294, 311–312.

27. Braudel, *Mediterranean*, I, 166–167.

28. Semple, *Geographic Environment*, 311–312; Fuller, *Operating Manual for Spaceship Earth*, 21. Webb sees such European expansions from the 1490s to the 1890s as a "boom" in the "Metropolis" precipitated by the challenges of the "Great Frontier." *The Great Frontier*, xiii, 8–15 and passim. Henning Henningsen argues that the history of seafaring peoples for various reasons has until very recently been neglected because of the predominance of the agrarian or land-oriented historian. "The Life of the Sailor Afloat and Ashore," in Hasslöf et al., *Ships and Shipyards*, 123, 124ff.
29. Mahan, *Influence of Sea Power*, 50.
30. Quoted in Frederic C. Lane, *Venice: A Maritime Republic* (Baltimore, 1973), 3–4, in which Lane compares this observation with the way a present-day New Yorker might view the villages of the downeast Maine coast.
31. J. H. Parry, *Trade and Dominion: The European Overseas Empires in the Eighteenth Century* (New York, 1971), 8–9, which lists three principal stages of maritime imperial growth—the "reconnaisance", the "collisions" and "industrial dominance."
32. Peter Mathias, *The First Industrial Nation: An Economic History of Britain, 1700–1914* (New York, 1969), 13.
33. Lane, "Republicanism," 521.
34. Thomas C. Cochran, "The Business Revolution," *The American Historical Review*, LXXIX, no. 5 (December 1974), 1466, in which his example is the early national northeastern United States and its seaports.
35. Lane, "Republicanism," 522, also 524.
36. Semple, *Geographic Environment*, 327.
37. See Anderson, *Ulysses Factor*, 32.
38. For example, mutinies in thalassocratic navies have been conspicuously rare, the major ones being those in the British navy at Spithead and the Nore in 1797. See G. J. Marcus, *The Age of Nelson* (New York, 1971), 82–101. Continental navies have suffered by contrast, with major mutinies in several modern fleets between 1905 and 1921: Russia (1905, 1917, 1921), Austria-Hungary (1918) and Germany (1917, 1918). See Richard Hough, *The Potemkin Mutiny* (New York, 1961); Daniel Horn, *The German Naval Mutinies of World War I* (New Brunswick, N.J., 1969); Anthony Sokol, *The Imperial and Royal Austro-Hungarian Navy* (Annapolis, 1968), 132–133; and Paul Avrich, *Kronstadt 1921* (Princeton, 1970), especially 54, 65.
39. The best study of the seamen of Britain, the longest lasting thalassocracy, remains Michael Lewis, *The Navy of Britain: A Historical Portrait* (London, 1948), 287–332. But see John Masefield, *Sea Life in Nelson's Time*, 3rd ed. (Annapolis, 1972) and Henry Baynham, *Before the Mast: Naval Ratings of the 19th Century* (London, 1971).
40. For the ancient world (when slaves were however used occasionally on merchant vessels), see Lionel Casson, *Ships and Seamanship in the Ancient World* (Princeton, 1971), 322–328. For Pisan, Genoese and Venetian war galleys, on which latter the presence of slaves was illegal, see Lane, *Venice*, 45, 133, 165. Whereas manpower shortages led continental states like Spain, France and Turkey to use slaves, convicts, debtors and prisoners, such occasionally hard-pressed thalassocracies as Venice and Britain resorted to the less extreme measures of conscription and impressment respectively. For Spain and France,

see Braudel, *Mediterranean*, I, 139 and Paul W. Bamford, *Fighting Ships and Prisons: The Mediterranean Galleys of France in the Age of Louis XIV* (Minneapolis, 1973).

41. Samuel Eliot Morison, *The European Discovery of America: The Southern Voyages, 1492-1616* (New York, 1974), viii; ". . . these paladins of discovery never had enough," though Morison makes no distinction between the nations these navigators served. On exploration, in addition to Anderson, *Ulysses Factor*, see J. H. Parry, *The Age of Reconnaisance* (New York, 1964), 19, and Albert Bettex (tr. by Daphne Woodward), *The Discovery of the World* (London, 1960), 374.

42. Mathias observes that a state with a purely agrarian economy can never aspire to industrial greatness due to its lack of surplus capital needed for "productive investment." *First Industrial Nation*, 5-6.

43. For the religion of the Renaissance seafarers, see Morison, *Southern Voyages*, 165, and J. R. Hale, *Renaissance Exploration* (New York, 1968), 90. "Shipboard religion" was a simple mixture of practical prayer, superstition and sea custom, with evangelicalism being left to land-lubbing clerics and missionaries. Morison, *The European Discovery of America: The Northern Voyages, A.D. 500-1600* (New York, 1971), 142-143.

44. The development *and spread* of a phonetic alphabet followed directly from the need to keep records of mercantile transactions. The seafaring Phoenicians deserve primary credit for the diffusion of the alphabet throughout the ancient world and no small share of the credit for its development. Sabatino Moscati (tr. by Alastair Hamilton), *The World of the Phoenicians* (New York, 1965), 88ff. Also Hans Jensen (tr. by George Unwin), *Sign, Symbol and Script: An Account of Man's Effort to Write* (New York, 1969).

45. See especially William McFee, *The Law of the Sea* (New York, 1950) and Pitman B. Potter, *The Freedom of the Seas in History, Law, and Politics* (London, 1924).

46. One example may be found in Harold D. Langley, *Social Reform in the United States Navy, 1798-1862* (Urbana, Ill., 1967), which traces a period during which the American navy was profoundly influenced by the British example.

47. For the Venetian and Dutch examples of attempts of wealthy thalassocracies to relieve the plight of their poor, see Lane, *Venice*, 108-109, 271, 331-334 and K. H. D. Haley, *The Dutch in the Seventeenth Century* (London, 1972), 152-165.

48. See, for the British navy, Christopher Lloyd, *The Navy and the Slave Trade* (New York, 1949); W. E. F. Ward, *The Royal Navy and the Slavers* (London, 1969) and J. B. Kelly, *Britain and the Persian Gulf, 1795-1880* (Oxford, 1968).

49. Examples may be found in H. A. Omerod, *Piracy in the Ancient World* (1924; Chicago, 1967); Grace Fox, *British Admirals and Chinese Pirates, 1832-1869* (London, 1940); and Harry Miller, *Pirates of the Far East* (London, 1970).

50. A concise account of the British experience is Neville Williams, *Contraband Cargoes: Seven Centuries of Smuggling* (London, 1959).

51. R. S. Allison, *Sea Diseases* (London, 1943) treats the history of British naval medicine, which led the world in the eighteenth and nineteenth centuries. Geoffrey Barraclough, *An Introduction to Contemporary History* (New York, 1964), 57-58 and passim, observes that the nineteenth-century empires

hastened their own demise by introducing modern European medicine, diet and education among their new theretofore primitive colonial peoples.

52. John U. Nef, *War and Human Progress: An Essay on the Rise of Industrial Civilization* (New York, 1963), 84–88, 328 and passim.

53. Peter Burke, *Culture and Society in Renaissance Italy, 1420–1540* (London, 1972), 292. One particular predominant mode of artistic expression over another was not relevant. For example, on the superiority of painting over literature in seventeenth-century Holland, see J. L. Price, *Culture and Society in the Dutch Republic during the 17th Century* (London, 1974).

54. Braudel, *Mediterranean*, II, 728–729; David Maland, *Culture and Society in Seventeenth-Century France* (New York, 1970), 12–13, and 158: "Unlike the merchants of Holland and England, few French merchants had an exclusive interest in commerce and industry for its own rewards. On the contrary they were enthralled by the grandeur of aristocratic life. . . ."

55. For a few but telling examples, the many Aegean thinkers who gravitated to Athens, the Florentine Petrarch and the German Dürer to Venice, the Frenchmen Descartes to Holland and Voltaire to England. Beyond the period of this essay, in the twentieth century scholar émigrés have fled totalitarian states to the security of the American political and intellectual sphere.

56. Semple, *Geographic Environment*, 294, the example in this particular discussion being France.

57. Ibid., 333. Semple wrote this after acknowledging that her mentor Ratzel had been heavily influenced by the social Darwinism of Herbert Spencer which she rejected, vi–vii. Says Toynbee, *A Study of History*, 114: inasmuch as "a harsh environment is not inimical to civilization but rather the contrary, are we warrented in formulating the further proposition that the stimulus towards civilization grows stronger in proportion as the environment grows more difficult?" But he accepts continental geographic stimuli equally if not more than the oceanic, a view shared by Frederick Jackson Turner in his case study of the United States. "The Significance of the Frontier in American History" (essay presented before the American Historical Association, 1893).

58. A. E. Taylor, for example, explains away Plato's account of the Atlantid kings' having had "the services of a highly-developed engineering science" as "a very characteristic Platonic touch." *A Commentary on Plato's Timaeus* (Oxford, 1928), 50–51. But in fact all thalassocracies tended to excel in such endeavors. A more charitable assessment is by Benjamin Jowett, *Plato: Timaeus* (Indianapolis, 1949), xxii–xxiii.

59. A recent example is J. V. Luce, *Lost Atlantis: New Light on an Old Legend* (New York, 1969).

60. *Critias*, 120d–121a. *The Dialogues of Plato*, 4 vols., 4th ed. (tr. by B. Jowett), III (Oxford, 1953), 804. The allusion to lack of interest in property, like Cassiodorus' description of the sixth-century A.D. Venetians cited above, would seem to describe an aversion to the landed wealth and extravagance that attends great continental aristocracies like those of contemporary Egypt and Persia. See Thucydides on Athens, below, in the present essay.

61. Ibid., 114c–121c, 797–804.

62. Ibid., 121a–121b, 804.

63. Isocrates, *On the Peace*, 102–103 in *Isocrates*, 3 vols. (tr. by George Norlin), II (New York, 1929), 73. In the *Panathenaicus*, 115–116 (II, 443, 445), Isocrates

uses Sparta and Athens to make a clear distinction between the two types of powers: "... a land-power is fostered by order and sobriety and discipline and other like qualities, [while] a sea-power is ... augmented by ... the crafts which have to do with the building of ships and by men who are able to row them," unpropertied men that Isocrates disparaged; by their entrance "into the state the order and discipline of the former polity would be relaxed," an obvious reference to the liberalizing effect of seamen.

64. Thucydides, *Peloponnesian War*, 15–16. This view has been rejected by Chester G. Starr, "The Myth of the Minoan Thalassocracy," *Historia*, III (1954–55), 282–291. This period is known to archeologists as the Late Minoan I–II.

65. H. G. Wells (rev. by Raymond Postgate and G. P. Wells), *The Outline of History* (London, 1971), 161. For the full artistic dimension, see Reynold Higgins, *Minoan and Mycenaean Art* (New York, 1967), which concludes (190) that Minoan Crete "was one of the really great civilizations of antiquity." Also, R. W. Hutchinson, *Prehistoric Crete* (Baltimore, 1962), 123–136, especially the reference to H. A. Groenewegen-Frankfort's analysis (132) of dynamic form in Minoan art; and Semple, *Geographic Environment*, 415–416. The absence of idols and religious monuments suggests also a very loose religious establishment.

66. The vast literature on the Minoans dates from A. J. Evans' pioneering work, *The Palace of Minos at Knossos*, 5 vols. (1921–35). The key recent work is Colin Renfrew, *The Emergence of Civilisation: The Cyclades and the Aegean in the Third Millenium B.C.* (London, 1972), but see also the shorter work by Sinclair Hood, *The Minoans: The Story of Bronze Age Crete* (New York, 1971).

67. See Luce, *Lost Atlantis*; H. E. L. Mellersh, *The Destruction of Knossos: The Rise and Fall of Minoan Crete* (New York, 1970); and D. L. Page, *The Santorini Volcano and the Desolation of Crete* (London, 1971).

68. William H. McNeill, *The Rise of the West: A History of the Human Community* (Chicago, 1963), 97–99.

69. Most of the above works agree on this, but see also George E. Mylonas, *Mycenae and the Mycenaean Age* (Princeton, 1966).

70. Jack M. Sasson, "Canaanite Maritime Involvement in the Second Millenium B.C.," *Journal of the American Oriental Society*, 86, no. 2 (1966), 128, an otherwise excellent treatise.

71. Moscati, *World of the Phoenicians*, 43, summarizes general scholarly opinion that the Phoenicians had "few aspirations to great art, while the craftsmanship displays a combination of influences and often lacks coherence and stylistic tradition." See also William Culican, *The First Merchant Venturers: The Ancient Levant in History and Commerce* (New York, 1966).

72. J. Holland Rose, *Man and the Sea: Stages in Maritime and Human Progress* (Cambridge, Eng., 1935), 38. Samuel Tolkowsky, *They Took to the Sea: A Historical Survey of Jewish Maritime Activities* (New York, 1964), is grossly overstated.

73. See Miller, *The Thalassocracies*, for a recent treatment of these pre-Athenian thalassocracies. Of Samos, Semple perhaps exaggerates its role as "a center of Ionian manners, luxury, art, science and culture, the seat of the first great thalassocracy or sea-power after that of Cretan Minos, a distributing point of commerce and colonies." *Geographic Environment*, 417.

74. Russell Meiggs, *The Athenian Empire* (Oxford, 1972), 273.

75. A contemporary example was the anonymous "old oligarch" who damned the economic aspects of Athenian command of the sea: "As for the cities on the mainland which are ruled by Athens the large are controlled by fear and the small by need. For there is no city which does not need to export or import; and these things it will not be able to do unless it accepts the bidding of the power that rules the sea." Quoted in ibid., 266. Wells is a typical modern example, calling Pericles "a great demagogue" albeit "perhaps the most honest of demagogues." But Wells cannot deny the overall Athenian contribution to mankind; with Judea, Athens had begun "a moral and an intellectual process in mankind, an appeal to righteousness and . . . to the truth. . . . Slowly, more and more, men apprehended the reality of human brotherhood, the needlessness of wars and cruelties and oppression, the possibilities of a common purpose for the whole of our kind. In every generation thereafter there is evidence of men seeking for that better order to which they feel our world must come." *The Outline of History*, 274, 277, 286.

76. Thucydides, *Peloponnesian War*, 118–119. Other translators reject Warner's "an education of Greece" in favor of "the school of Hellas." See Stringfellow Barr, *The Will of Zeus* (New York, 1961), 162, which also says (165) that Athens was a school in two senses, inciting "other Hellenic states by her contagious example to achieve democratic freedom" and acting as "a learning polis" for her own citizens. The Athenian teachings spread across the Mediterranean with Greek colonies and traders. See John Boardman, *The Greeks Overseas* (Baltimore, 1964).

77. Quoted in Meiggs, *The Athenian Empire*, 399. Of the Athenian chronicler Xenophon, Warner has said that he "never realized, as Pericles had realized long before, that the institutions and manners of Sparta and indeed of all landlocked and constricted states were already out of date, inadequate for the present and for the future." Introduction to Xenophon (tr. by Rex Warner), *History of My Times (Hellenica)* (Baltimore, 1966), 11.

78. Toynbee, *A Study of History*, 116.

79. McNeill, *Rise of the West*, 257–258. But also, observes George Sarton, "In spite of its national eminence and its cosmopolitanism, the Athenian culture remained original and spontaneous." *A History of Science: Ancient Science Through the Golden Age of Greece* (Cambridge, Mass., 1952; New York, 1964), 224. Athenian religion was superficial to the point of being almost secular.

80. T. B. L. Webster, *Athenian Culture and Society* (Berkeley, 1973), 265.

81. Thucydides, *Peloponnesian War*, 51. "Thus Athens, because of the very variety of her experience, is a far more modern state than you are." 52. Says Semple, "The ancient philosophers of Greece were awake to the deep-rooted differences between an inland and a maritime city, especially in respect to receptivity of ideas, activity of intellect, and affinity of culture." *Geographic Environment*, 281.

82. M. Cary, *A History of the Greek World from 323 to 146 B.C.* (London, 1932, 1972), 242. See also Irwin Loeb Merker, "Studies in Sea-Power in the Eastern Mediterranean in the Century Following the Death of Alexander." (Unpublished Ph.D. dissertation, Princeton University, 1958).

83. Cary, *Greek World*, 307.

84. Ulrich Wilcken (tr. by G. C. Richards), *Alexander the Great* (New York, 1957), 260.
85. John Marlowe, *The Golden Age of Alexandria* (London, 1971), 237.
86. Good overviews are B. H. Warmington, *Carthage* (London, 1960) and Emeline Richardson, *The Etruscans: Their Art and Civilization* (Chicago, 1964).
87. Lionel Casson, *The Ancient Mariners* (New York, 1959), 157.
88. Ch. I of J. H. Thiel, *Studies on the History of Roman Sea-Power in Republican Times* (Amsterdam, 1946), 1-31.
89. Ibid., 4 (original in italics), 28, 20, 22 (italics original), 18. Still useful is Frederick William Clark, *The Influence of Sea-Power on the History of the Roman Republic* (Menasha, Wis., 1915).
90. Thiel, *Roman Sea-Power*, 16.
91. F. E. Adcock, *The Roman Art of War Under the Republic* (New York, 1963), 30.
92. Ibid., 37.
93. Dietmar Kienast, *Untersuchen zu den Kriegsflotten der Romischen Kaiserzeit* (Bonn, 1966). The standard work on the navy in the Empire remains Chester G. Starr, *The Roman Imperial Navy, 31 B.C.-A.D. 324* (Ithaca, 1941).
94. M. P. Charlesworth, *Trade-Routes and Commerce of the Roman Empire* (1926; New York, 1970), 8-13. Rose shrewdly observes that such safety upon the sea encouraged Christian missionaries like St. Paul to spread the gospel. *Man and the Sea*, 48-49.
95. [Flavius Renatus] Vegetius (U.S. Army War College tr.), *Military Institutions of the Romans*, Book V (mimeographed, 1928), 83.
96. Charles Diehl (tr. by Naomi Walford), *Byzantium: Greatness and Decline* (New Brunswick, N.J., 1957), 25, and 105-107 for a brief overview of the considerable artistic contributions of Constantinople, a major land and sea entrepôt.
97. McNeill, *Rise of the West*, 451.
98. See Archibald R. Lewis, *The Northern Seas: Shipping and Commerce in Northern Europe, A.D. 300-1100* (Princeton, 1958). For the Vikings, see Gwyn Jones, *A History of the Vikings* (New York, 1968). The impressive beginnings of English maritime growth may be glimpsed in Lewis' "England as an Atlantic Maritime Power, 1100-1350 A.D." (Unpublished paper, Medieval Academy Annual Meeting, Chapel Hill, N.C., 1971).
99. See Archibald R. Lewis, *Naval Power and Trade in the Mediterranean, A.D. 500-1100* (Princeton, 1951); Ekkehard Eickhoff, *Seekrieg und Seepolitik zwischen Byzantinischer und Arabischer Hegemonie (650-1040)* (Berlin, 1966); Aly Mohamed Fahmy, *Muslim Sea-Power in the Eastern Mediterranean From the Seventh to the Tenth Century A.D.* (London, 1950); and George Fadlo Hourani, *Arab Seafaring in the Indian Ocean in Ancient and Early Medieval Times* (Princeton, 1951).
100. McNeill, *Rise of the West*, 361ff.
101. Ibid., 524ff.; Jung-pang Lo, "The Emergence of China as a Sea Power during the Late Sung and Early Yüan Periods," *Far Eastern Quarterly*, XIV (August 1955), 489-503; C. P. FitzGerald, *The Southern Expansion of the Chinese People* (London, 1972); but especially Joseph Needham, *Science and Civilisation in China, IV, Physics and Physical Technology, Part III: Civil Engineering and Nautics* (Cambridge, Eng., 1971), 379ff.

102. George Woolley Spencer, "The Kingdom of Srivijaya and its Extensions," Ch. 5 of "Royal Leadership and Imperial Conquest in Medieval South India: The Naval Expedition of Rajendra Chola I, c. 1025 A.D." (unpublished Ph.D. dissertation, University of California, Berkeley, 1967), 146, 126, 149, 173. The dissertation disspells notions about Chola Indian attempts at maritime and naval supremacy in the Bay of Bengal (see 95–96). Shorter-lived Indonesian seaborne kingdoms of Majapahit and Malacca replaced Srivijaya in the fourteenth and fifteenth centuries.

103. Herbert Rosinski (ed. by Richard P. Stebbins), *Power and Human Destiny* (New York, 1965), 82–86.

104. Lane emphasizes the republican character of these city-states as being their "most distinctive and significant aspect . . . republicanism gave to the civilization of Italy from the thirteenth through the sixteenth centuries its distinctive quality. . . ." "Republicanism," 520, also 535 and passim.

105. Burke, *Culture and Society in Renaissance Italy*, 284, a work introduced by an excellent historiographical essay on the difficult problem of cultural creativity as revealed in the Italian Renaissance, 3–21.

106. McNeill, *Rise of the West*, 546.

107. Lane, *Venice*, 1. See also William H. McNeill, *Venice: The Hinge of Europe, 1081–1797* (Chicago, 1974), especially 46ff., 90ff.

108. D. S. Chambers, *The Imperial Age of Venice, 1380–1580* (London, 1970), 21, 33.

109. Oliver Logan, *Culture and Society in Venice, 1470–1790* (London, 1972), 68, 38 and passim. For Florence, see Ferdinand Schevill, *History of Florence from the Founding of the City through the Renaissance* (New York, 1936, 1961), which describes Florence as a merchant society, xv, 288, 291–292, 307–309 and passim; also Lane, "Republicanism," 526–529.

110. Spanish Catalonia had relied on Genoese and Pisan maritime immigrants as early as the twelfth century for its "thriving maritime economy." Braudel, *Mediterranean*, I, 147. For the oceanic period, see J. H. Elliott, *Imperial Spain, 1469–1716* (New York, 1963).

111. Braudel, *Mediterranean*, II, 728; Michael E. Mallett, *The Florentine Galleys in the Fifteenth Century* (Oxford, 1967).

112. Braudel, *Mediterranean*, II, 728, which cites Hermann Hefele's 1929 essay "Zum Begriff der Renaissance" in *Historisches Jahrbuch*.

113. Schevill, *Florence*, 287ff.; Lane, *Venice*, 170, 201, 245. For Florentine and Venetian culture in the years after thalassocracy, see Eric Cochrane, *Florence in the Forgotten Centuries, 1527–1800* (Chicago, 1973) and McNeill, *Venice, the Hinge*, 155ff.

114. Braudel, *Mediterranean*, I, 290–292, 312; II, 728–729, 1242.

115. Columbus was a Genoese, Amerigo Vespucci a Florentine whose importance has however been thoroughly demolished by Morison, *Southern Voyages*, 294–297.

116. Braudel, *Mediterranean*, II, 899, suggests an explanation in "that any economic recession leaves a certain amount of money lying idle in the coffers of the rich: the prodigal spending of this capital, for lack of investment openings, might produce a brilliant civilization lasting years or even decades."

117. Parry, *Discovery of the Sea*, 264.

118. Mahan, *Influence of Sea Power*, 50–52.
119. Parry, *Discovery of the Sea*, 88–90, 92–93, 95–97, 124.
120. J. H. Elliott, *The Old World and the New, 1492-1650* (Cambridge, Eng., 1970), 8.
121. J. H. Parry, *The Spanish Seaborne Empire* (London, 1966), 25. See also C. R. Boxer, *The Portuguese Seaborne Empire, 1415-1825* (New York, 1969).
122. See Andrew C. Hess, "The Evolution of the Ottoman Seaborne Empire in the Age of Discoveries, 1453–1525," *The American Historical Review*, LXXV, no. 7 (Dec. 1970), 1892–1919; Arthur J. Marder, "From Jimmu Tennō to Perry— Sea Power in Early Japanese History," *The American Historical Review*, LI, no. 1 (Oct. 1945), 1–34; Toynbee, *A Study of History*, 407ff.; and the above works on China. These states were seaborne only in the narrow strategic sense as a subordinate element of continental preoccupations, the spirit in which C. R. Boxer refers to Portugal's "century-long thalassocracy in the Indian Ocean." "Portugal's *Drang nach Osten*," *The American Historical Review*, LXXV, no. 6 (Oct. 1970), 1686.
123. Braudel, *Mediterranean*, I, 138, 146–147; II, 725–727, 755.
124. Ibid., II, 1241.
125. Ibid., II, 1242.
126. Ibid., II, 1240 (perhaps even as late as 1680); Elliott, *Old World and the New*, passim.
127. Elliott, *Old World and the New*, 100, 77–78 and passim. This frontier theme expands and improves upon Webb's *Great Frontier*.
128. Nef, *War and Human Progress*, 213; McNeill, *Rise of the West*, 580–581, 679; F. L. Carsten, "Introduction: The Age of Louis XIV," *The New Cambridge Modern History, V: The Ascendancy of France, 1648-88* (Cambridge, Eng., 1961), 17–18. For Denmark, see Charles W. Petersen, "England and Danish Naval Strategy in the Seventeenth Century" (unpublished Ph.D. dissertation, University of Maine, 1975). For Sweden's maritime efforts in the age of Gustavus Adolphus and Charles XII, see *Svenska Flottans Historia*, 3 vols. (n.p., 1942–45), I, II. For Holland, England and France, see below.
129. J. H. Huizinga (tr. by Arnold J. Pomerans), *Dutch Civilisation in the Seventeenth Century, and other essays* (London, 1968), 16, 35, 60, 112. In his essay on the Romans, Dutchman Thiel admires his native countrymen's respect for the sea: "fear of the hostile element, sometimes even hatred and aversion form the background of the pious [Dutch] confidence in the mercy of God, which keeps those sailors on their legs. . . ." *Roman Sea-Power*, 3. Mahan describes the Dutch state as "republican in name, [allowing] large scope to personal freedom and enterprise, and the centers of power were in the great cities." *Influence of Sea Power*, 55.
130. Charles Wilson, *The Dutch Republic, and the Civilisation of the Seventeenth Century* (New York, 1968), 18.
131. Haley, *Dutch in the Seventeenth Century*, 194.
132. From the Introduction (xxv) to C. R. Boxer, *The Dutch Seaborne Empire* (New York, 1965), which see for a full treatment of the imperial achievement. Says Haley, "The Netherlands was a great cultural crossroads, a market for the buying and selling of ideas and books as well as corn, spices, sugar and fish." *Dutch in the Seventeenth Century*, 192. The basis for the achievement in

science "was seaborne trade, shipping and navigation" stemming from the need for longer and lower-cost voyages. Wilson, *Dutch Republic,* 92. In addition to the above sources which also treat the artistic dimension of Dutch culture, see Kenneth Clark, *Civilisation* (New York, 1969), 193–201. An excellent overview remains Pieter Geyl, *The Netherlands in the Seventeenth Century,* 2 vols., rev. and enl. ed. (New York, 1961–64).

133. See Boxer, *Dutch Seaborne Empire.* Mahan ascribes the colonies' weakness to their "essentially commercial" character; as "mere commercial dependencies upon the mother-country" the colonies never developed into politically stable units. *Influence of Sea Power,* 57.

134. Haley, *Dutch in the Seventeenth Century,* 192–193.

135. Huizinga, *Dutch Civilisation,* 101–103.

136. Wilson, *Dutch Republic,* 230, 242.

137. G. S. Graham, *The Politics of Naval Supremacy* (Cambridge, Eng., 1965), 20–22.

138. John Knox Laughton, *Studies in Naval History* (London, 1887), 39. J. Lough, "France Under Louis XIV," in *The New Cambridge Modern History,* V, 243, also stresses the English example. For Colbert's mercantile origins, see Jean-Louis Bourgeon, *Les Colbert avant Colbert: Destin d'une famille marchande* (Paris, 1973).

139. Voltaire (tr. by Martyn P. Pollack), *The Age of Louis XIV* (London, 1966), 333. Mahan admires the "adventureous temper" of the French people, but mourns their economic temper which "leads them to seek [profit] by thrift, economy, hoarding . . ., [and] when excessive prudence or financial timidity becomes a national trait, it must hamper the expansion of commerce and of the nation's shipping." *Influence of Sea Power,* 53–54.

140. "The fact is that seafaring was . . . a peripheral [activity] in France" because of two land frontiers; "The French kings from Louis XI through Henry II were far from indifferent to seaports and mariners, but their series of wars . . . made the army their chief concern." By the time of Louis XIV, the North American continent as an imperial frontier had thus come to be dominated by Britain. Morison, *The Northern Voyages,* 273.

141. ". . . the careful administration of Colbert, whilst able to build and equip ships in a manner equal to those of any country in Europe, was not able to improvise, in a similar way, a race of men who could be compared with the sailors of Holland." Laughton, *Studies in Naval History,* 53. See also Mahan, *Influence of Sea Power,* 36, 54–55, and Donald George Pilgrim, "The Uses and Limitations of French Naval Power in the Reign of Louis XIV: the Administration of the Marquis de Seignelay, 1683–1690" (unpublished Ph.D. dissertation, Brown University, 1969), 5, 34n.55, 407-408, 421ff. and passim. Says Pilgrim (27), "A continental perspective seems to have influenced France's diplomatic historians as well as her diplomats." Still standard is Charles Woolsey Cole, *Colbert and a Century of French Mercantilism,* 2 vols. (New York, 1939), which concludes (II, 549–550) that Colbertism was inevitable after 1661 though not necessarily in the Colbert manner of organization and effectiveness. See also Eugene Asher, *The Resistance to the Maritime Classes: the Survival of Feudalism in the France of Colbert* (Los Angeles, 1960). Lacking an adequate natural maritime manpower base, France had early turned to impressing debtors and other convicts. See Bamford, *Fighting Ships and Prisons.*

142. Mahan, *Influence of Sea Power*, 70–71.
143. Maland, *Culture and Society in Seventeenth-Century France*, 10–11, 15, 236. See also David Ogg, "The Achievements of France in Art, Thought and Literature," in *The New Cambridge Modern History*, V, 248ff., and Wells, *The Outline of History*, 691.
144. Maland, *Culture and Society in Seventeenth-Century France*, 12–13, 157–158. See n.54 above.
145. Nef, *War and Human Progress*, 148–151, 214–218.
146. Seeking a strategic alternative to a big battle fleet, the French repeatedly turned to commerce raiding (*guerre de course*), while the navy was charged with administering the distant colonies (even after the creation of the Colonial Office in 1892). See Raoul Castex, *Synthèse de la Guerre Sous-marine de Pontchartrain à Tirpitz* (Paris, 1920) and other works by this prolific writer-admiral; Theodore Ropp, "The Development of a Modern Navy: French Naval Policy, 1871–1914" (unpublished Ph.D. dissertation, Harvard University, 1937), summarized in his "Continental Doctrines of Sea Power" in Edward Mead Earle, ed., *Makers of Modern Strategy* (Princeton, 1943), 446–456; and Ronald Chalmers Hood III, "The Crisis of Civil-Naval Relations in France, 1924–1939: No *Concorde* Between Them" (unpublished M.A. thesis, University of Maine, 1972).
147. Louis B. Wright, *Gold, Glory and the Gospel: The Adventurous Lives and Times of the Renaissance Explorers* (New York, 1970), 336ff.
148. Nef, *War and Human Progress*, 328. This passage refers to the period 1792–1815 and includes the young United States, but it may be applied to the British thalassocracy over three centuries. See also Mahan, *Influence of Sea Power*, 29–30, 36–37.
149. Rose, *Man and the Sea*, 99, which sees (99–100) the promise of overseas English economic and intellectual expansion in Sir Thomas More's *Utopia* (1516).
150. Mahan, *Influence of Sea Power*, 36–37.
151. McNeill, *Rise of the West*, 733–734. Ralph Davis, in contrasting eighteenth-century England and France, sees England as having a shorter period of state regulation of industry, a more diversified religious life, and "a more open society. Social classes did not mix freely, but above the level of the poorest they did mix a little." *The Rise of the Atlantic Economies* (Ithaca, 1973), 309–310. See also Mathias, *First Industrial Nation*. The openness of British society, like that of preceding thalassocracies, invited residency by foreign intellectuals, notably from Enlightenment France.
152. See, for instance, Clark, *Civilisation*, 245–289.
153. Rosinski, *Power and Human Destiny*, 162, and my *Command of the Sea*, 321ff.
154. McNeill, *Rise of the West*, 734–739. McNeill also (734) attacks the notion, largely Marxist, that a capitalistic-liberal society like Britain's necessarily leads to industrial greatness.
155. Ibid., 739.
156. For Germany at sea, for example see Theodore Ropp, "German Seapower: A Study in Failure," in A. M. J. Hyatt, ed., *Dreadnought to Polaris: Maritime Strategy Since Mahan* (Toronto, 1973), 12–18; several recent unpublished dissertations: Albert Harding Ganz, "The Role of the Imperial German Navy in Colonial Affairs" (Ohio State University, 1972); Patrick James Kelly, "The

Naval Policy of Imperial Germany, 1900–1914" (Georgetown University, 1970); and Keith W. Bird, "Officers and Republic: The German Navy and Politics" (Duke University, 1972); also, Jost Dülffer, *Weimar, Hitler und die Marine: Reichspolitik und Flottenbau, 1920–1939* (Düsseldorf, 1973) and Walter Ansel, *Hitler and the Middle Sea* (Durham, N.C., 1972), 484.

157. Nef, *War and Human Progress*, 328; Rose, *Man and the Sea*, 101–104; Parry, *Age of Reconnaisance*, 283; Mahan, *Influence of Sea Power*, 38–39, 57–58; McNeill, *Rise of the West*, 670.

158. Lane, *Venice*, 377.

159. See John J. Kelly, Jr., "The Struggle for American Seaborne Independence as Viewed by John Adams" (unpublished Ph.D. dissertation, University of Maine, 1973). This period includes the administration of George Washington (1789–1797), during which John Adams was Vice President, and which was largely shaped by the entrepreneurial policies of the Secretary of the Treasury, New Yorker Alexander Hamilton (1789–1795).

160. See Cochran, "Business Revolution," 1449–1466, in which he supports America's common ties with England and contrasts the unfavorable conditions for entrepreneurial development in France, Germany and the Low Countries during the early Industrial Revolution (1461). Also, J. Wade Caruthers, "The Influence of Maritime Trade in Early American Development: 1750–1830," *The American Neptune*, XXIX, no. 3 (July 1969), 199–210.

161. Van Wyck Brooks, *The Flowering of New England, 1815–1865* ([New York,] 1940) and Ralph H. Gabriel, *The Course of American Democratic Thought* (New York, 1940, 1956), especially 70ff.

162. Benjamin W. Labaree in Labaree, Robert G. Albion and William A. Baker, *New England and the Sea* (Middletown, Conn., 1973), 43–44.

163. Woodward, "Age of Reinterpretation"; Ellen Churchill Semple, *American History and its Geographic Conditions* (Boston, 1903), 114ff. and 226ff. Anglo-American strategic maritime similarities are traced in my essay, "The British Strategic Inheritance in American Naval Policy, 1775–1975," in Benjamin W. Labaree, ed., *The Atlantic World of Robert G. Albion* (Middletown, Conn., 1975), pp. 169–194, 238–248.

164. E. Berkeley Tompkins, for example, synthesizes the causes of American imperialism in strong thalassocratic terms: industrial growth linked to overseas colonial markets; the example of European imperialism; social Darwinism; "Manifest Destiny"; the theoretical closing of the continental frontier; and Protestant missionaries. *Anti-Imperialism in the United States: The Great Debate, 1890–1920* (Philadelphia, 1970), 4, 5, 7, 291. A frustrated Mahan tried to fabricate the preconditions for American thalassocracy: "The instinct for commerce, bold enterprise in the pursuit of gain, and a keen scent for the trials that lead to it, all exist; and if there be in the future any fields calling for colonization, it cannot be doubted that Americans will carry to them all their inherited aptitude for self-government and independent growth." *Influence of Sea Power*, 57–58. See my "Thalassocratic Determinism of Captain Mahan," *University of Maine 1971 Seminar Proceedings*, 77–85, also Milton Plesur, *America's Outward Thrust: Approaches to Foreign Affairs, 1865–1890* (DeKalb, Ill., 1971). Wells sees in twentieth-century America the hope of man, "In that continent will be the greater opportunity, the greater wealth, the greater freedom of mind." *The Outline of History*, 863.

165. R. J. Kerner, *The Urge to the Sea: The Course of Russian History* (Berkeley, 1926), and Semple, *Geographic Environment*, 344–345. See, for the United States, George Rogers Taylor, *The Transportation Revolution, 1815–1860* (New York, 1951) and Howard B. Schonberger, *Transportation to the Seaboard: The "Communications Revolution" and American Foreign Policy, 1860–1890* (Westport, Conn., 1971).

166. See my *Command of the Sea* for both nations, and for Russia especially Robert Waring Herrick, *Soviet Naval Strategy: Fifty Years of Theory and Practice* (Annapolis, 1968) and David Fairhall, *Russian Sea Power* (Boston, 1971).

167. McNeill, *Rise of the West*, 726ff., especially 806–807, and Page Smith, *The Historian and History* (New York, 1964), especially 58–65, 98–109. One difficulty is the dearth of universal historians since World War II; "the syntheses of Fernand Braudel and William H. McNeill are the only serious attempts at presenting global history . . . within a well-defined theoretical framework." Niels Steensgaard, "Universal History for Our Times," *The Journal of Modern History*, 45, no. 1 (March 1973), 72.

168. Fuller, *Operating Manual for Spaceship Earth* and his many other writings all related to this theme. Another example is Ritchie Calder, *Man and the Cosmos: The Nature of Science Today* (New York, 1968).

169. John F. Kennedy, "The Space Challenge: High National Priority," address at Rice University, Sept. 13, 1962, reprinted in Lester M. Hirsch, ed., *Man and Space* (New York, 1966), 139, 140. For "global cosmopolitanism," see McNeill, *Rise of the West*, 726ff.

170. James G. Allen, "Historical and Philosophical Background of the Space Age," in William Frye, ed., *Impact of Space Exploration on Society*, VIII of American Astronautical Society Science and Technology Series (Washington, 1966), 16. See also Sidney Hyman, "Man on the Moon: The Columbian Dilemma," in Eugene Rabinowitch and Richard S. Lewis, eds., *Man on the Moon: The Impact of Science, Technology, and International Cooperation* (New York, 1969), 39–52.

171. Arthur C. Clarke, *The Promise of Space* (New York, 1968), 352.

172. Arthur C. Clarke, "Beyond Apollo," in Neil Armstrong et al., *First on the Moon* (Boston, 1970), 414. A science fact-fiction writer, Clarke typifies many professional scientists writing in the same vein. See, for example astrophysicist-exobiologist Carl Sagan, *The Cosmic Connection: An Extraterrestrial Perspective* (Garden City, 1973), 67–69, 156–157, and mathematician Freeman Dyson, "Human Consequences of the Exploration of Space," in Rabinowitch and Lewis, *Man on the Moon*, 13–27.

173. Toynbee, *A Study of History*, 47.

3

CAPTAIN MAHAN,
THALASSOCRATIC DETERMINIST

As noted in the preceding chapter, the American naval officer/ historian Alfred Thayer Mahan was a major proponent of "sea power," a concept he defined and applied to the conduct of particular nations in the past. In fact, he was the first—and most famous—modern historian to treat the subject as a strategic-historical overview. His motives, however, were purely nationalistic—a U.S. Navy captain who used history to advocate the maritime, naval, and colonial expansion of his own country (he retired as captain in 1896 and was promoted to rear admiral on the retired list ten years later).

In his zeal, Mahan overstated his case and fell into the philosophical trap of explaining history largely in terms of this single factor—sea power—a pitfall faced by any historian who ranges across the broad sweep of the past. This chapter demonstrates how, intellectually, he succumbed to this temptation. It is therefore a brief guide to understanding Mahan's limitations as a historian.

Originally entitled "The Thalassocratic Determinism of Captain Mahan," the chapter was delivered as a scholarly paper in a session on "The Philosopher-Historians in the Golden Age of Naval Thought" at a conference at the University of Maine in 1971 and is reprinted from the proceedings (Orono, Maine: University of Maine Printing Office, 1972). It was written before the publication of Mahan's papers, edited by Robert Seager II and Doris Maguire (1975) and of the definitive biography by Seager (1977). The papers contain a wealth of nuggets which reveal details of Mahan's thalassocratic determinism. For a list

Captain Mahan, Thalassocratic Determinist

of Mahan's writings, see John B. Hattendorf and Lynn G. Hattendorf, compilers, *A Bibliography of The Works of Alfred Thayer Mahan* (Newport, R.I.: Naval War College Press, 1986).

The late nineteenth century witnessed not only a golden age of naval thought but a period of extraordinary historical theorizing. Western society of the last century needed "the cohesion of history," according to Page Smith, in order "to counteract the centrifugal forces of the modern world—the rise of industrialism . . .; the growth of skepticism; the development of extreme self-consciousness in groups and classes within nations, and national self-consciousness itself."[1] Thus, in 1890, when Alfred Thayer Mahan suddenly appeared as the first major synthesizer of naval history, important theories regarding the nature of man were being formulated or expounded by such of his contemporaries in philosophy and history as, in Europe, Jacob Burkhardt, Wilhelm Dilthey, Friedrich Nietzsche, Sigmund Freud, J. B. Bury and Oswald Spengler, and, in the United States, Henry Adams, Frederick Jackson Turner and Charles A. Beard.[2] Mahan is not generally classed among these men as an original thinker but rather as a leading naval analyst and historical strategist.[3] His chief fame rested upon his naval-maritime theories in which, in his own words, he tried "to show how the control of the sea, commercial and military, had been an object powerful [enough] to influence the policies of nations; and equally a mighty factor in the success or failure of those policies."[4] Hence contemporary and subsequent historians referred to Mahan for his authority on naval matters.

In fact, however, Mahan was a major historical philosopher who deserves recognition alongside the above thinkers, especially next to his fellow countrymen Turner and Beard, whose major works and thus published theories he antedated and whose subject matter, American history, was much narrower than Mahan's. Sweeping across modern European history, Mahan presented his main theories in 1890—in the first chapter of his most important book, *The Influence of Sea Power upon History, 1630–1783*.[5] Turner's environmental thesis—the impact of the frontier on the shaping of the American character—did not appear until three years later, and Beard's economic interpretation of the American Constitution was not published for another twenty years.[6] The influence of Mahan's ideas upon these two men, if indeed there was any, would be difficult to assess, but the fact remains that Mahan utilized strong environmental and economic arguments to develop his own historical thesis. Mahan believed, namely, that certain "natural" or "general conditions" existed within seafaring individuals throughout history "that either are essential to or powerfully affect the greatness of a nation upon the sea."[7] That Mahan's original philosophical ideas failed to create a special historical school as did those of Turner and Beard was largely due to the greater attention given his naval theories and to their general decline after 1914. Mahan's death that year, at the onset of twentieth-century technological warfare, deprived him of the chance to revise and perhaps update both his historical and strategic theories. Mahan has therefore been generally regarded as a pre-1914 thinker, irrelevant both to modern strategy and general philosophical history.[8]

Mahan became a thalassocratic determinist. Traditionally, to repeat, the Greek word *thalassa*, meaning sea, has been used to describe an empire built upon dominion over the seas and adjacent lands. This was Thucydides' definition of the thalassocracy of Minos in second millenium B.C. Crete,[9] while recently C. R. Boxer used the term for a similar description of the sixteenth-century Portuguese seaborne empire as a "century-long thalassocracy in the Indian Ocean."[10] Mahan clearly understood this simple understanding of a maritime empire and thus formulated his economic basis for such an empire in the term "sea power." To Mahan, sea power required three things: production, shipping (commercial and naval) and colonies.[11] This was the formula his popularizers and readers quickly understood and applied at the turn of the century. Thus his central thesis was economic as well as being political and military: the influence of sea power—that is, production, shipping and colonies—upon the European powers from 1500 to his own day. And this theme has been the one traditionally understood as the basis for Mahan's maritime determinism.

In order to prove his economic theories, however, Mahan believed he needed to account for some cultural or even inherited trait as the force of any true national enterprise upon the sea. So he developed his environmental thesis, the search for what he termed variously as "the character and pursuits of the people," their "natural impulses" or "natural bias" or "bent" or "attitude" or "social sentiment."[12] The effort placed Mahan in the naturalistic tradition of the late-Enlightenment German cultural philosopher Johann Gottfried von Herder and on a level with his own younger contemporary Frederick Jackson Turner.[13] Whereas Turner viewed the American character as being molded by its special relationship with primitive natural conditions on the frontier, Mahan sought to explain the character of all seagoing peoples in terms of "the natural growth of a people's industries and its [the people's] tendencies to seek adventure and gain by way of the sea. . . ."[14] This environmental argument thus complemented the economic one and became the philosophical foundation for Mahan's larger thalassocratic determinism.

The intellectual challenge of adequately formulating and proving his environmental thesis was perhaps beyond Mahan's mental abilities. He even tried to excuse himself from the task by finding it "necessary to avoid a tendency to over-philosophizing, to confine attention to obvious and immediate causes and their plain results, without prying too far beneath the surface for remote and ultimate influence."[15] He therefore tried to simplify his presentation by creating a list of six "natural conditions" necessary for the development of a vital thalassocracy.[16] Three of the conditions described the ideal geography and topography for a coastal people for overseas travel, and three formulated the cultural factors behind a seagoing people and a government sympathetic to the creation of sea power.[17] Though convenient, this listing of ingredients such as the number of merchant seamen, the habits of the people, and the capitalistic nature of their government did not solve the central problem posed by Mahan: which came first, the so-called natural inclinations of a people toward the sea, or the enlightened direction of their government in creating the conditions for sea power? Put another way, were a people thalassocratic by nature or by governmental fiat? Mahan could not be sure, even though he analyzed—albeit superficially—five European maritime powers before testing his thesis on his own country. Despite the shortcomings of his case studies, Mahan was the first to apply systematically a common yardstick to measure the success or failure of these early nation-states on the sea.

The sixteenth-century Iberian powers Spain and Portugal Mahan considered failures at developing sea power. Both peoples "had many great qualities; they were bold, enterprising, temperate, patient of suffering, enthusiastic, and gifted with intense national feeling," but both governments were too Medieval and feudalistic to develop seaborne empires naturally. Of the Spanish government, which also ruled Portugal from 1580 to 1640, Mahan said it was "in many ways such as to cramp and blight a free and healthy development of private enterprise; but the character of a great people breaks through or shapes the character of its government, and it can hardly be doubted that had the bent of the people been toward trade, the action of government would have been drawn into the same current."[18] Mahan saw the same lack of maritime character in Portugal, whose fortunes "followed the same downward path; although foremost in the beginning of the race for development at sea, she fell utterly behind."[19] Recent scholarship has generally supported Mahan's conclusions about Spain and Portugal, providing the necessary evidence and adding another important line of inquiry. That is, how did overseas exploration, colonization and trade affect the character of the Iberian people? Very little before 1650, according to J. H. Elliott, for the discovery of a new world "constituted a challenge to a whole body of traditional assumptions, beliefs and attitudes."[20] So entrenched were these social and religious customs that especially Spain resisted the intellectual opportunities raised by her new empire; concludes J. H. Parry, "Discovery is not necessarily a social catalyst, nor the possession of empire a harbinger of change."[21] Lacking the true nature of seagoing peoples, Spain and Portugal failed Mahan's test to qualify as sea powers and thus as true thalassocracies.

Mahan held out greater prospect for the seventeenth-century Dutch, although the land frontier to Louis XIV's France ultimately doomed the United Provinces. Mahan envisioned thalassocratic greatness in the Netherlands; "the State was republican in name, allowed large scope to personal freedom and enterprise, and the centers of power were in the great cities."[22] He compared the Dutch with the English. Both had "wisdom and uprightness" and "were by nature businessmen, traders, producers, negotiators. . . ."[23] The Dutch thus had had two elements of sea power: production and shipping. They also had colonies, which however were so "purely commercial in character" that they never grew politically. Said Mahan: "This placid satisfaction, with gain alone, unaccompanied by political ambition, tended, like the despotism of France and Spain, to keep the colonies mere commercial dependencies upon the mother-country, and so killed the natural principle of growth."[24] Why the Netherlands failed ultimately has been explained thoroughly by the recent work of Charles Boxer, namely, because the Dutch were simply overextended—given their small population and heavy trade commitments in continental Europe and throughout their far-flung colonies.[25] The test of cultural vitality the Dutch did pass; J. H. Plumb observes that "science and philosophy were both stimulated by the great extension of human knowledge that the geographies brought about. The transocean trade brought more than a profit: it made windows into the mind."[26] Because of the eventual military and thus economic collapse of the seventeenth-century Netherlands, Mahan minimized the social-cultural nature of the Dutch character and thus would not accord them full thalassocratic status.

Where the Dutch failed, in Mahan's view, the English succeeded. The geographic insularity of eighteenth-century Great Britain produced all the prerequisites for the ideal thalassocracy. The English people had everything, especially enduring naval might, with which to succeed upon the sea: "Their many wants, combined with their restless activity and other conditions that favored maritime enterprise, led her people abroad. . . . Their needs and genius made them merchants and colonists, then manufacturers and producers; and between products and colonies shipping is the inevitable link. So their sea power grew."[27] Where the English character surpassed the Dutch was revealed in the colonist. The English overseas settler "naturally and readily settles down in his new country, identifies his interests with it . . . [and] at once and instinctively seeks to develop the resources of the new country in the broadest sense."[28] The thalassocratic nature of the British nation has been so widely accepted, with no small credit due to Mahan's own historical narratives, that it has become almost common knowledge.

France, in the maritime interlude under Louis XIV and his famous minister Colbert, presented Mahan with greater difficulties. "France," he said, "has a fine country, an industrious people, an admirable position"; the French people had an "adventureous temper, which risks what is has [in order] to gain more, [and] has much in common with the adventureous spirit that conquers worlds for commerce."[29] But he observed that the French aristocracy belittled commercial pursuits, French merchants were excessively timid and cautious, and the French peasant was too wedded to the soil to feel any urge to the sea.[30] Mahan therefore concluded that France had not the natural maritime character for building a geniune thalassocracy. This lack, however, he said was overcome by the excellent French sense of system as exemplified by Colbert, who almost singlehandedly—with the blessing of Louis—created French sea power, that is, production, shipping and colonies, in the 1660s and 1670s. Lacking a maritime nature, France's growth upon the sea depended upon the sustained program of a thalassocratic-minded government. "Such growth," observed Mahan, ". . . was forced, and depended upon the endurance of absolute power which watched over it."[31] Colbert's death in 1683 and French naval reverses shortly thereafter doomed the French attempt to achieve maritime grandeur, although Mahan refused to allow himself to believe that the French had nothing at all maritime in their basic nature.[32]

Recent scholarship is no more certain about the French maritime aptitude than was Mahan. Gerald S. Graham in his *Politics of Naval Supremacy* accords France a "great maritime inheritance" from being "in many respects an amphibious state" with "a race of superb seamen" on two coastlines that "inevitably stimulated the maritime and commercial ardour of French governments." But even though the French were "a versatile people (who seemed to possess a special aptitude for colonization)," Graham seems to conclude that the French failed abroad not because of the lack of a thalassocratic nature but because of France's military commitment on the continent and thus the lack of geographic insularity.[33] Donald M. Schurman in his *Education of a Navy* sees Mahan's fascination with France both in terms of his ingenious technique of using French sources to write basically pro-British history and because of Mahan's reliance on the writings of the French-Swiss military analyst Henri Jomini, whose teachings were advanced at West Point by Mahan's father, Dennis Hart Mahan.[34] I would go further and suggest that Mahan's initial—and therefore most profound—impressions about commercial and maritime policy came from a French history text, Henri Martin's *History of France,* which became a "kind of introductory primer" for Mahan in his study of Colbert's system, which he greatly admired.[35]

Mahan's final case study—the United States—plainly failed. Though the Americans of the 1890s could draw upon their English maritime inheritance, Mahan conceded that they had only the production prerequisite for sea power. Their commercial and naval shipping were inadequate, and they had no colonies at all.[36] Had he stopped there, his analysis of the American character would have had very little impact within his own country. Instead, he abandoned his environmental critique and launched into propaganda. Suggesting that the American urge to the sea might only be latent from colonial times, he implored his countrymen and their government to create a thalassocratic empire—clearly following the French example.[37] When the people and their Congress were thus awakened, he believed, American sea power would emerge: "The instinct for commerce, bold enterprise in the pursuit of gain, and a keen scent for the trails that lead to it, all exist; and if there be in the future any fields calling for colonization, it cannot be doubted that Americans will carry to them all their inherited aptitude for self-government and independent growth."[38] Clearly, Mahan's notion of Americans being good colonizers was wishful thinking; he could not have selected a more stay-at-home people for his optimism in this respect.[39] If the United States created an empire after 1898 and again—by revisionist standards—after 1945, it was not following Mahan's formula. Americans rather favored protectorates which, rather than colonizing them, they have Americanized by imposing their own political, social and commercial values on the indigenous peoples.

His American chauvinism aside, Mahan did ask significant questions about the nature of seafaring man. He was the first even to suggest that certain preconditions exist for the emergence of a thalassocratic state; his failure lies in the superficial manner in which he tried to determine these preconditions. His essay of 1890 deserves a place at least alongside Turner's essay of 1893 which, like Mahan's, overstated its case and was thus left open to subsequent attack.[40] Indeed, the overstatement of the impact of the role of seafaring on the nature of man has led to some recent cautious reaction. For example, the Israeli historian of ancient seafarers, Zvi Herman, recently prefaced his book with the statement: "The doctrine preached by this school, which one might call thalassologists, or people who place stress on the significance of the sea and the part it has played in the crystallization of human history is, without doubt, too extreme." Then, he adds, almost contradicting himself, "Nevertheless, we should recognize that this doctrine rests on pretty solid foundations."[41]

Indeed, what I would call the thalassocratic factor—first introduced by Mahan—is finally being elevated to the forefront of historical scholarship. Not only have the national histories by such scholars as Plumb, Parry, Graham and Boxer begun to test it, but it has found expression in such general histories as J. H. Elliott's *The Old World and the New* and William H. McNeill's *The Rise of the West*.[42] Also, implicit in Mahan's discussion, the question arose regarding the impact of exploration on men, recently the subject of a special study by J. R. L. Anderson entitled *The Ulysses Factor: The Exploring Instinct in Man*.[43] Unfortunately, Mahan's economic and military theories of sea power and navies overshadowed completely the thalassocratic determinism upon which they were based. Those theories aside, the study of man's relationship with the sea—and its effect on his culture and institutions—can proceed, and scholars can build upon the crude beginnings of Mahan's essay.

NOTES

1. Page Smith, *The Historian and History* (New York, 1964), p. 56.
2. In 1890, when Mahan turned 50 years old, Burkhardt was 72, Dilthey 57, Nietzsche 46, Freud 34, Bury 29, Spengler 20, Adams 52, Turner and Beard 26. Charles Darwin had been dead eight years, Leopold von Ranke four.
3. Alongside John and Philip Colomb, John Laughton, Herbert Richmond and Julian Corbett, all British. See D. M. Schurman, *The Education of a Navy: The Development of British Naval Strategic Thought, 1867–1914* (Chicago, 1965). Ch. 4 covers Mahan. For a brief strategic discussion of "historical" and "mate-

rial" schools of strategic thought, see Captain S. W. Roskill, *The Strategy of Sea Power* (London, 1962), pp. 101ff., especially 109 and 139.

4. Captain A. T. Mahan, USN (Ret.), *From Sail to Steam: Recollections of Naval Life* (New York, 1907), p. 283.

5. (Boston, 1890), Ch. I, "Discussion of the Elements of Sea Power," pp. 25–89. Hereinafter referred to as Mahan, "Essay."

6. Frederick Jackson Turner, "The Significance of the Frontier in American History" (Paper read before the American Historical Association, 1893); Charles A. Beard, *An Economic Interpretation of the Constitution of the United States* (New York, 1913).

7. Mahan, *Influence of Sea Power upon History*, p. 23. The word "natural" appears in this context throughout the essay.

8. For criticism of his naval thought, see Margaret Tuttle Sprout, "Mahan: Evangelist of Sea Power," in E. M. Earle, ed. *Makers of Modern Strategy* (Princeton, 1943), pp. 444–45; Theodore Ropp, "Continental Doctrines of Sea Power," in the same anthology, pp. 454–56; William F. Livezey, *Mahan on Sea Power* (Norman, Okla., 1947), pp. 274–75, 299; Gerald S. Graham, *The Politics of Naval Supremacy* (Cambridge, Eng., 1965), pp. 3, 5, 28–29; Clark G. Reynolds, "Sea Power in the Twentieth Century," *Journal Royal United Services Institute,* CXI (May 1966), 132–36. For changes in historical thought from 1914, see Page Smith, *The Historian and History,* pp. 66ff.

9. *The Peloponnesian War,* I, 4.

10. "Portugal's *Drang nach Osten,*" *American Historical Review,* LXXV. No. 6 (October 1970), 1686.

11. Mahan, "Essay," p. 28.

12. Catchwords repeated throughout the essay.

13. See Smith, *The Historian and History,* pp. 110–11.

14. Mahan, "Essay," p. 82.

15. Ibid., p. 58.

16. Ibid., p. 28.

17. Ibid., pp. 29ff.

18. Ibid., pp. 50–51.

19. Ibid., pp. 51–52.

20. J. H. Elliott, *The Old World and the New, 1492–1650* (Cambridge, England, 1970), p. 8. Though speaking generally of Europe, Elliott's emphasis is on the Latin-speaking peoples. Also see Elliott's *Imperial Spain, 1469–1716* (New York, 1963).

21. J. H. Parry, *The Spanish Seaborne Empire* (London, 1966), p. 25. See also Parry's *The Age of Reconnaisance* (New York, 1963) and Charles R. Boxer, *The Portuguese Seaborne Empire: 1415–1825* (New York, 1969).

22. Mahan, "Essay," p. 55.

23. Ibid., pp. 52–53.

24. Ibid., p. 57.

25. C. R. Boxer, *The Dutch Seaborne Empire* (New York, 1965).

26. Plumb's introduction to ibid., p. xxv. For the artistic dimension, see Kenneth Clark, *Civilisation* (New York, 1969), pp. 193–201.

27. Mahan, "Essay," pp. 36–37.

28. Ibid., p. 57.

29. Ibid., p. 53.
30. Ibid., pp. 36, 54–55.
31. Ibid., pp. 70–71.
32. See ibid., pp. 76, 77, 80. The maritime zeal of the French people he felt was shown under Choiseul in the 1760s and again after the initial naval defeats during the Revolution.
33. G. S. Graham, *Politics of Naval Supremacy*, pp. 20–22.
34. D. M. Schurman, *Education of a Navy*, pp. 69–70.
35. A. T. Mahan, *From Sail to Steam*, pp. 280–82.
36. Mahan, "Essay," p. 84.
37. Ibid., pp. 38–39, 76, 88.
38. Ibid., pp. 57–58.
39. In the vast literature on American isolationism, see Selig Adler, *The Isolationist Impulse* (New York, 1957) and, more recently, E. Berkeley Tompkins, *Anti-Imperialism in the United States: The Great Debate, 1890–1920* (Philadelphia, 1970).
40. Some of Richard Hofstadter's constructive comments about the frontier thesis are applicable to correcting the excesses of Mahan. *The Progressive Historians* (New York, 1969), pp. 159–64.
41. Zvi Herman, (tr. by Len Ortzen), *Peoples, Seas and Ships* (London, 1964), pp. xii–xiii.
42. (Chicago, 1963), pp. 97–99, in which McNeill contrasts thalassocratic Minos with the militaristic Hittites.
43. (New York, 1970). See also Albert Bettex (tr. by Daphne Woodward), *The Discovery of the World* (London, 1960), p. 374.

4

AMERICA AS A THALASSOCRACY—
AN OVERVIEW

Mahan's tentativeness about the United States becoming a full-fledged thalassocracy, as discussed in the previous chapter, influenced me until I examined America afresh, in the context of previous modern maritime models, notably Great Britain. In so doing, I discovered that military and naval historians writing after Mahan's death (in 1914) had also been deeply affected by Mahan's history and so-called principles of sea power.

The result of my examinations are this chapter and the following one, which reinterpret the United States in strategic terms not only before and during Mahan's lifetime but in the years since. This chapter is a popularly written, and somewhat personal, summary of my findings. It was commissioned as a bicentennial appreciation of America upon the sea by the United States Naval Institute and appeared in the Institute's *Proceedings* for July 1976.

The next chapter explains, among other things, how American military and naval historians were affected by Mahan's view of history in the decades following his death.

The dependence of the American people upon the sea—to utilize it for political and economic power and even for cultural growth—has been so dramatically demonstrated throughout our history as to make the image of the republic incomprehensible without it. Without American activity upon

---

the waters, alone and with allies, how different the course of American and world history would have been—from the crossing of the Pilgrim fathers on the *Mayflower* to the recovery of the Apollo astronauts. Without the central role of New England shippers within the British Empire, how could the colonies have achieved their economic importance and urge toward independence? Without the support of the French fleet at Yorktown, how could the Continental Army have ever culminated that movement to independence? Without British and American naval patrols, how could the Monroe Doctrine have been enforced and thus American continental expansion accomplished without European interference? Could the Union have been preserved without Lincoln's blockade? Would the nation have connected the oceans without Mahan's ideas and Teddy Roosevelt's Panama Canal and modern fleet? Indeed, would not the continent of Europe have become German and all East Asia Japanese without the intervention of the U.S. Navy and the sealifted American ground, air, and logistical forces in two World Wars?

On this auspicious occasion, the 200th anniversary of the birth of the Republic, an accounting is in order of just how important has been the role of sea power in the history of the United States. So let us now recall the monumental importance of our citizenry upon the sea in those 200 years. With even a superficial appreciation of this historical force, we can then better understand the vital role of the United States in the progress of the human race.

THE PAX

O Sea! Thou Sea!
What nations thou hast framed,
What thoughts inspired,
What fires kindled high.

Americans have always prided themselves on being "peace" loving people, preferring to compete in the economic, technological, diplomatic, and even athletic arenas rather than resorting to armed force. However, as every student of politics and international affairs knows all too well, each nation has unique interests and goals, and no two nations have ever coexisted in complete harmony. International tensions between national powers have been the norm since the beginning of recorded history. Simply put, when these tensions are in balance, we have peace, and when the balance of the powers is upset, we have war. Thus, this country has sought peace but has occasionally had to resort to war with the shifting power balance.

The United States has always been a virtual island nation in relation to the other great power centers of the world. The surrounding waters have been less a natural barrier than the medium upon which naval power has kept the nation isolated from the wars between those other powers, or have been used to project American power hence on the few occasions when the country felt forced to fight. Had the nation been threatened by the presence of large standing armies perpetually on its land borders, no doubt it would have relied on a similar large army of its own. But such a threat has never existed (save for the Civil War period), whereas America's enemies and potential enemies have always threatened from the sea, making the Navy the first line of defense.

The U.S. Navy, however, has not always possessed the capability to provide this defense unilaterally. Indeed, throughout most of American history, it was never strong enough to protect America's political interests alone. It thus had to act instead as a diplomatic tool in political situations dominated by other and larger navies. Consequently, from 1778 to 1781, isolated Continental Navy vessels and squadrons merely augmented the allied French fleet which then insured ultimate independence. In the 1798–1800 quasi-war, the new and small U.S. Navy supported the Royal Navy against the French, then reversed its support again from 1812 to 1815. During most of the *Pax Britannica*, Britain's Royal Navy deterred general war in Europe and policed the sea lanes of the world; the latter function was shared by the U.S. Navy. And between 1898 and 1947, a general if sometimes shaky Anglo-American naval partnership existed in peace and war between the two powers until, finally, after 1947 general deterrence became chiefly an American responsibility, giving the period its strategic identity as the *Pax Americana*.

The great goal of American foreign policy throughout these two centuries has remained constant: the maintenance of international law and order throughout the world so that the American people might go about their business peaceably. Consequently, the nation has insisted on freedom of the seas for trade; on the suppression of illegal use of the sea by pirates, slave traders, and smugglers; and in the present century on the preservation of the global balance of power to insure political and economic stability. Throughout, sea power has been the primary means of enforcing this American interpretation of international maritime law.

The desire for economic independence by maritime New England most directly triggered the War for Independence (1775–1783) from the British Empire, the quasi-war with France (1798–1800), the Barbary Wars (1801–1815), and the unsuccessful attempt at neutrality during the Napoleonic Wars (1803–1814). Though the application of sea power in each of these affairs resulted in defeats as well as victories, the net effect was to convince Britain to drop her trade barriers and to accept free trade (completely by the 1850s) for which the United States and other nations had been fighting and bargaining.

Henceforth, the Anglo-American peoples shared a common interpretation of international law, although the United States refused any official tie and declined to endorse the general European Declaration of Paris (1856) on international law. The desire for free trade with Britain played a large part in America's decision to intervene in World War I following the German declaration of unrestricted submarine warfare on our shipping (1917). And freedom of the seas was a pillar of President Woodrow Wilson's Fourteen Points at the Versailles Conference (1919). In this case, the policy was aimed against the British. A strong Navy—able to deter or fight those nations which would deny us the free use of the sea—made such policies possible. Even when the British blocked our neutral trade with wartime Germany in the early years of both World Wars, we ultimately chose alliance with Britain against a common foe.

Illegal use of the oceans by violators of international law led to an active American role of naval policing, usually in concert with other Western nations. The Barbary pirates were only the first of many such maritime outlaws to feel the weight of American naval guns and cutlasses. The Navy was especially active between 1815 and 1860 in the Caribbean, Southeast Asian waters and the South Atlantic, where slave traders defied Western laws against the practice. Piracy still occurs, as shown in the *Mayaguez* incident (1975). Anti-smuggling work was the province of the Revenue Marine Service, later the U.S. Coast Guard. And adventurers who attempted to run the declared Union blockade in defiance of federal authority found themselves hunted and ultimately thwarted by four Union blockading squadrons (1861–1865). Warships of a growing Navy were the agents of enforcement of international law as interpreted by our government.

The desire of twentieth-century American governments to preserve the balance of power in Europe and Asia has led to several major unilateral American interpretations of international law. The interpretations have succeeded because of the Navy's ability to enforce them. Presidents Wilson and Franklin D. Roosevelt both regarded the possible collapse of Britain to the international outlaws of Germany as inimical to American interests—political, economic and moral—thus prompting our intervention in both World Wars. And all American Presidents from William McKinley (1897) to FDR (1945) regarded Japanese incursions into China with similar disdain, leading to the Open Door Policy (1900) and eventually to our unofficial participation in the Sino-Japanese War (1941), climaxed by Japan's treacherous and illegal attack on Pearl Harbor. In the thermonuclear Cold War, the United States interpreted the presence of Russian missiles in Cuba as a threat to the global balance of power and hence intervened with a naval blockade in 1962. In each such crisis, the U.S. Navy applied its sea power in concert with the sister services to affirm the American view of international law and order.

Indeed, since 1945, American sea power has grown beyond the confines of being a mere diplomatic tool in the shadow of the Royal Navy. It has become the prime deterrent global force as well as the most important police force in maintaining the worldwide *pax*. Before World War I, Theodore Roosevelt's battleship diplomacy helped deter the Germans in the Caribbean (Venezuela, 1903, and the "Big Stick") and Mediterranean (Morocco crisis, 1904–05 and 1913) and Japan in the Pacific (especially with the Great White Fleet, 1908), while American warships patrolled in such diverse places as the Yangtze River and the Virgin Islands. In the 1920s and 1930s, this policing was intensified, with the U.S. Fleet concentrated in the Pacific to deter Japan. When the United States allowed this deterrent and the patrol forces to deteriorate, Japan was encouraged to initiate the war against China in 1937.

World War II—won largely by superior allied fleets, sealifted armies and air forces, and seaborne logistics—destroyed not only the German and Japanese empires but proved equally fatal to those of Britain, France, and Holland. Into the resulting power vacuum stepped the United States, its fleets henceforth patrolling the oceans of the world and projecting American military power ashore in troubled coastal regions from Korea (1950) and Vietnam (1965) to Lebanon (1958) and the Dominican Republic (1965). This strategy, at first (1947) politically designed to "contain" global Communism, eventually matured into the Nixon Doctrine (1969) which formalized America's basic maritime strategy of controlling the sea and air of the Asian periphery while its allies provide the manpower. In so doing, American sea power acts as a deterrent to limited war in those regions and throughout the world.

The twentieth century also witnessed the first attempts at international arms control, wherein quantitative measures were established to limit arms production and inventories. Significantly, such yardsticks were not in armies or land weapons or in aircraft but in capital ships and seaborne weapon systems. The first halting attempts, at the Hague conferences of 1899 and 1907, considered several issues but focused mainly on the laws of naval warfare and neutrality— partly because of the contributions of Alfred Thayer Mahan of the U.S. Navy. The subsequent rise of American naval strength enabled the United States to convene the Washington Conference on the limitation of naval armaments in 1921–22. In this conference arms controls were established through battleship and carrier tonnage quotas, gun restrictions, and the freezing of the general strategic-geographic balance in the Pacific. This successful arrangement—due largely to Anglo-American-Japanese cooperation—was renewed at the London Conference of 1930 but then collapsed in 1936 with the rise of the uncooperative dictatorships. This experiment in arms control, in which the world's navies remained the object, set important precedents for the future—namely, the limitation of missile-launching submarines and their warheads.

America as a Thalassocracy

Throughout our history, the Navy has taken its major inspiration from the Royal Navy, although we have drawn on particular technical and tactical innovations of most of the other major navies. By the present century, when the Navy became a true oceangoing force, it shared with Britain a common concern over Germany and Japan, with both English-speaking navies developing doctrines, bases, and tactics with which to counter these two belligerent nations. Following the victory over them in World War II, the Anglo-American navies looked upon Soviet Russia as the common foe, and since Russia's naval strength lay in its submarine armada, Allied naval doctrines shifted to antisubmarine warfare. With the advent of nuclear weapons, however, the U.S. Navy developed its own nuclear war capability, matching that of the new U.S. Air Force after severe interservice battles in the late 1940s. During the 1950s, the Navy deployed nuclear warheads on carrier-based A3D Skywarriors and developed the Polaris submarine/missile system for deployment during the 1960s. Such seaborne mobility and range gave the Navy the ability to attack continental targets in Russia just as it had struck Japan in the closing months of World War II. The arms race of the years 1945–1962 gave the Russo-American Cold War its character, based on the belief that a thermonuclear exchange in a World War III was a viable military alternative.

The Cuban Missile Crisis of 1962 showed the folly of this assumption, the aftermath being the tacit recognition by the United States and the Soviet Union that such a general war would be counterproductive to both nations and indeed to the survival of industrial civilization. Nuclear weapon systems—America's being matched by similar Soviet systems—have since been regarded as a separate political category outside the normal conduct of military and naval operations, although the advent of tactical nuclear weapons like the ship-to-ship cruise missile has weakened distinctions between general and conventional war forces. Nevertheless, the tense thermonuclear arms race, or Cold War, reached its climax during the Cuban crisis. Since then, the Americans and Russians have sought to control these weapons, culminating in the Strategic Arms Limitation Treaty (SALT I) in 1972 with revisions during the present year. The Polaris/Poseidon/

Trident ICBM submarines and carrier-based, nuclear-armed attack planes have great mobility and range—unlike the Strategic Air Command's fixed, hardened missile silos and bomber bases—which add a unique dimension to Russia's strategic woes, since both require constant surveillance by the Russians. The sea-based nuclear deterrent therefore remains the key to the current arms balance as the ultimate form of military power projected from the sea. What third parties such as China, France, India, or Israel do in the nuclear arena only time will tell, but for the moment these American and Russian weapons are the yardstick for offensive deterrence and arms control and thus act as the strategic umbrella over the current *pax.*

Just as Soviet Russia developed its own strategic missile subs to match ours, it has also greatly expanded its surface fleet and antiship submarine attack forces for a conventional naval presence. In addition, it has taken a commanding lead in seagoing fishing factories and oceanographic research vessels and has significantly enlarged its merchant marine. Also, the Russians have supplied naval vessels and weapons to their allies and third world countries, just as has the United States. The U.S. Navy thus has had to counter all these activities by maintaining a viable strategic deterrent against the Soviet (and Chinese) interior and coast, including naval bases and airfields, against Russian surface forces and attack subs, and against Russia's commercial fleets. At the same time, we must look to our tactical fleet defenses in order to maintain our ability to control the open seas. This Russian naval and maritime growth has caused great concern and debate within American naval and political circles. At the very least, we need new ships and weapons to meet this vast and complex challenge. If we have them, the Navy should be able to maintain a credible conventional naval deterrent stance. For Russia lacks the doctrines, balanced fleet, overseas bases, at-sea logistics, major amphibious forces, and— above all—the *experience* to wage a conventional war on the oceans. Strong at sea, the U.S. Navy can deter Russia from strategic adventurism in this or any environment.

Needless to say, whenever deterrence and maritime policing— and the *pax*—have failed, war has followed, and therein sea power has assumed its most violent manifestations. America's cruisers and gunboats of the sailing era, though few in number, gave distinguished accounts of the nation's military prowess from 1775 to the 1830s in otherwise defensive wars—defensive of the coast and of commerce. Blockade, amphibious, and riverine operations in the Mexican War, the Arrow War in China, and Civil War (1846–1865) initiated the Navy into a limited offensive stance. The Spanish-American War (1898) began the modern offensive American Navy, projecting sea power and armies into Cuba, Puerto Rico, and the Philippines, then throughout the Caribbean (1901–1914), to North Africa and France (1917–1918, 1942–1945), into Russia (1918–1919), the Black Sea and Eastern Mediterranean (1919–1923), the Pacific islands (1942–1945), Korea (1950–1953), and South Vietnam (1965–1973). American "peacetime" aid went by sea to Western Europe on three occasions (1914–1917, 1939–1941, and 1946–1949) to shore up the Western democracies. These temporary measures resulted in war for America in the first two cases and in the latter in the country's first peacetime defensive alliance, the North Atlantic Treaty Organization (1949). NATO's communications and flanks ever since have been protected by American-centered naval forces.

By 1976, American sea power has extended the nation's peacetime policy of freedom of the seas over the entire globe, enabling general intercourse between all seagoing nations to continue unabated, save for local disputes such as the Anglo-Icelandic "cod war." If international agreements on laws to govern the sea finally emerge, or if new defensive treaties are enacted, it will still be American sea power which will enforce them. Even should new agreements not occur, or should they fail, the United States will unilaterally continue to use its sea power to enforce free exchange upon the sea. This is the legal right of any maritime nation—in our case the most powerful nation in history at that—much of whose livelihood is derived from the sea. Happily, most of the civilized world—which profits thereby—is grateful for it.

ECONOMIC GROWTH

And Sea! Our Sea!
From Cretan birth to Athens' main—
By star and wave
We learned to fly.

Sea power has been a vital ingredient of American economic growth in both overt and less obvious ways. Inasmuch as capitalism is based upon trade, the merchant and manufacturer have been interdependent foundation stones within the American economic framework and thus accountable for America's overall prosperity. Overseas trade has therefore been the basis of what has been somewhat inaccurately termed America's "empire." In the eighteenth century, Britain's colonies provided both the markets and sources of raw materials as part of a closed mercantilistic system, leading Mahan at the turn of the present century to the false conclusion that colonies therefore must provide an inevitable link in American sea power. But unlike the British, Americans have never been overseas colonizers, relying more so on trade, internal resources and domestic as well as global markets, so that their brief imperial experience in 1898 was unusual. The Philippines were hardly comparable to India, nor Puerto Rico to Transvaal. No, Americans have preferred their economic competition straight—by private enterprise, with a minimum of governmental interference, regulation, and taxes and therefore in a peacetime environment rather than under wartime controls.

Since the bulk of American foreign trade has been by sea, carried in private hulls and protected by government warships, shipbuilding and the carrying trade have provided major stimuli to the American economy. Competition has been fostered by the American belief in free enterprise and free trade, though protected by government tariffs and treaties and assisted by government subsidies and warship/weapon system contracts. The presence of Navy yards in certain coastal areas has also helped to stimulate the local economies in those regions. As for actual oceanic commodities, the fisheries (including for many years pelagic sealing and whaling) have been important down to the present, and offshore petroleum drilling and seabed mining have been recently added. This intimate interrelationship between the sea and economic power is older than the republic itself.

The North American colonies of Massachusetts and Virginia were originally founded by enterprising joint-stock companies following the English peace with Spain in the early seventeenth century. Joined by subsequent colonies, they soon became an integral part of England's sea-linked empire. By the next century the colonial seaports were prospering by direct and indirect trade with Europe, Africa, and the West Indies. This was so to the extent that especially maritime New England resented crown controls and eventually led the movement toward independence. These Northeastern merchants exhibited similar truculent economic independence under the new government, notably through the "Essex junto" and then the Hartford Convention which protested the economically adverse War of 1812. Allied with coastal New York and to a lesser extent with Philadelphia and Baltimore, all connected to the interior by rivers and canals, and championing the business-oriented politics of New York's Alexander Hamilton, maritime New England remained the undisputed center of American commerce for the first century of the republic.

American economic prosperity throughout the nineteenth century depended as much on the trade with Britain as America's strategic insularity rested on the good offices of the Royal Navy. In the 1790s, Anglo-American trade developed first in the Caribbean and spread outward from there, interrupted only by the American embargo of 1807–1809 and War of 1812. Both events were largely American economic protests against British policies. These temporary dislocations in fact stimulated internal American industrial development which then helped to magnify postwar trade. The Anglo-American seaborne exchange so prospered after 1815 that commercial reciprocity was accomplished by 1830, open free trade by the 1850s.

Economic surpluses that were thereby accumulated through overseas trade went into federal internal improvements—notably canals and turnpikes—which linked up the Great Lakes and the interior with the seaboard, especially as the river steamboat became available. In this way the Erie Canal alone from 1825 stimulated "seaport" development at Buffalo, Detroit, and Chicago, whence goods traveled via the canal to Albany and New York City. Jefferson's purchase of the vast Louisiana Territory in 1803 made New Orleans a seaport rivaled only by New York and Boston and turned the Mississippi River into a virtual second coastline along with the Atlantic Coast. The great river and its tributaries opened up the Midwestern interior to port developments as far northeastward as Cincinnati and Pittsburgh, southeastward to Nashville, westward to St. Louis, and northward to St. Paul, while Chicago was linked by canal to the Mississippi in 1848. By such river, coastal, and overseas traffic, culminating in the magnificent clipper ships, the volume of American merchant traffic nearly equalled Britain's by the 1850s, with an American near-monopoly over the key Liverpool-New York run.

Geographical expansion south and west during the nineteenth century—made possible partly by North America's physical isolation behind the Atlantic Ocean—had equally dramatic economic consequences. New Orleans was joined by Mobile, Pensacola, Galveston, and Houston in the Gulf maritime traffic as Spain's American empire waned. Diplomatic compromise with Britain resulted in the acquisition of port sites in the Pacific Northwest, and victory in the Mexican War added the tremendously important San Francisco Bay to the American economic complex. The gold rush—by " '49ers" on board swift clippers around Cape Horn to California—and American commercial inroads into China, Japan, and other Far Eastern markets stimulated the rise of the United States as a Pacific and thus global maritime power. The use and eventual acquisition of Samoa and Hawaii provided commercial way stations across this great new highway, creating the economic basis for the United States making the Pacific a sphere of economic and thus political influence by the end of the century. All these trade routes also served as avenues of immigration to America, to which foreign peoples came largely in hope of economic gain.

American economic power at sea suffered drastically for a number of reasons from the middle of the 1850s. The new railroads provided a more convenient and profitable means of freight and passenger transportation throughout the expanding continental interior, so much so that it successfully outstripped shipping from the beginning of the railroad boom in the 1850s. The financial panic of 1857 had a disastrous impact on world shipping, not least that of America whose shipbuilders and shipowners failed to respond with vessels more utilitarian than the speedy but small capacity clippers. Furthermore, the steam revolution had gone to sea; the British government subsidized private steamship development, while the United States did not. British steam thus rapidly bypassed American sail power. The American seagoing merchants instead shifted away from oceanic to coastal trade, thus surrendering commercial leadership at sea to the British. Confederate raiders hastened this shift, sinking many Yankee vessels and frightening even more to shift to neutral British registry because of lower wartime insurance rates and the protection of the Royal Navy. Finally, after the Civil War, an irate Congress refused to allow such errant merchantmen to return to the flag.

The American merchant marine never recovered from such reverses. The emergence of the modern dynamic American industrial state in the post-Civil War period did, however, lead foreign powers to eagerly seek trade with the United States, regardless of the registry of the carrier. And where the American merchant marine failed, high protective tariffs guaranteed enormous profits to entrepreneurs. Also, American business investments and close trade associations abroad provided powerful stimuli to American political and military interventions after 1898 in the Caribbean (Cuba and the Spanish-American War; Roosevelt Corollary to the Monroe Doctrine; Taft's Dollar Diplomacy), in China (Boxer Rebellion; Open Door Policy; Yangtze gunboat patrol), and in Europe (World War I).

Despite U.S. difficulties in oceanic trade, the nation's simultaneous territorial expansion reflected another key economic role of the sea. The acquisition of new coastal areas on the Gulf and Pacific Coasts broadened the range of the coastal fisheries from the cod of New England to the tuna of the Pacific Northwest, with increasing varieties of fish in between. Also, American dietary habits changed as lobster, shrimp, and abalone were added to the established shellfish tastes for oysters and clams late in the nineteenth century, and whales for oil and seals for hides sent Yankee fishermen into the colder waters of both major oceans. The purchase of Alaska and the Aleutian Islands from Russia in 1867 greatly extended the range of American commercial activities in the Pacific. They became even more pronounced following the discovery of Alaskan gold in 1897. The next year Hawaii became an American territory, giving the country not only a major new seaport at Pearl Harbor but the pineapple industry.

These sudden developments, combined with the receipt of the Philippines and Guam as spoils of war from Spain and the annexation of Wake, all in 1898, made the matter of interocean commerce especially urgent for economic as well as military reasons. The narrow isthmus of Panama had for centuries served as a terminus for ships transferring goods between oceans, but the hopes for a canal intensified to the point where Teddy Roosevelt supported the successful Panamanian revolution against ruling Colombia there in 1903 and then leased the Canal Zone. Construction was completed in 1914, just in time for the United States to exert international control over this vital new waterway at the outbreak of World War I. That control has continued down to the present, and though the railroad and commercial trucking eventually became preferred transcontinental carriers, the Panama Canal remains a major link in international trade. The Canal Zone and the states of Alaska and Hawaii are linked to the mainland states by ships—coastal merchantmen, warships, and Coast Guard vessels. Small wonder, then, that these acquisitions helped the republic to become the commercial colossus of the Western Hemisphere.

The United States, like the major industrial powers of Europe, sought economic self-sufficiency, a chief cause of the neo-imperialism fervor and arms race of the pre-World War I era. One such area of vital strategic importance was warship construction, leading to the Domestic Materials Act of 1886 in which Congress ruled that henceforth all American naval vessels were to be manufactured entirely in this country (a rigidity that has since been slightly relaxed). The burgeoning new American steel industry thus obtained rich government contracts for the new steel Navy. Such a policy did not follow for merchant shipping because of general and ever-growing British friendliness. But as the United States drifted into World War I, such independent American hulls became badly needed. Consequently, through the concerted efforts of President Wilson, Congress created the United States Shipping Board in 1916 and the Emergency Fleet Corporation in 1917 to provide additional vessels. The former achieved enormous success. After the war, the government disposed of these merchantmen and by the Washington Conference agreements froze its capital warship inventory; both deeds severely hampered the shipbuilding industry. The Merchant Marine Acts of 1928 and 1936 and the Public Works Administration of the New Deal stimulated some merchant and warship construction respectively, but neither could reverse the disastrous economic effects of the great depression.

In the 1940s, the previous wartime and postwar pattern was repeated. The rapid construction of the great two-ocean Navy which relieved Britain and Russia, helped to defeat the German and Italian navies and to destroy the Japanese Navy in World War II. This was both a tribute to American industrial capacity and a powerful stimulus to the American maritime economy. Postwar demobilization hindered naval construction until the Korean War, during which the Navy began to be revitalized. In the merchant marine, the Maritime Commission, created in 1936, achieved the epitome of American mass production techniques by directing—largely through the efforts of industrialist Henry J. Kaiser—the rapid creation of the wartime emergency merchant fleet of Liberty and Victory ships. But again, after the war, these were disposed of outright by sale or gift to destitute foreign nations or were mothballed.

The Merchant Marine Act of 1970 resumed federal efforts to stimulate American merchant ship construction, although longshoremen and mariners alike demand such high wages and safety measures by international standards that foreign registry continues to be economically preferable. The modern Navy continues to be a vital force in the American economy, but the costs of contemporary weapon systems have tended to stifle competition in favor of superrich corporate conglomerates.

Interrelationships between sea power and American economic growth remain strong as ever today. As the world's major industrial power, the United States—by example, trade, and foreign aid—has made the sea into a greater arena of international intercourse and "peacetime" rivalry than ever before. By example, the American maritime success has prompted Soviet Russia to compete in virtually every sphere of oceanic activity. Also, small countries with coastlines have learned to insist upon their fair share of the maritime economic pie—the basic cause of their insistence upon the 200-mile limit to protect their fisheries and ocean bed resources. Such nations which are oil rich have been major determinants in the current energy crisis, especially as they affect an oil-driven U.S. Navy and merchant marine and those of our allies.

By trade, the United States has come to depend upon the cheaper registries of minor states like Panama, Liberia, and Greece to carry American goods. Furthermore, American seaborne manufacturers and foodstuffs led to the recovery of postwar Western Europe and Japan until each has become a major economic competitor. By the same token, any country that urgently desires American economic and technological aid—now virtually *every nation in the world*— knows that such goods must come by sea.

Hence, the age-old American policy of free enterprise upon the seas has been not only a bedrock of American prosperity but has evolved into a prerequisite for the future economic survival of the human race.

SCIENCE AND TECHNOLOGY

The tides of time, the winds of change
By Venice to the Dutch—
Salt air for life;
Eternal vigor, power complete.

The sciences of the sea and seafaring have always had far-reaching implications for human knowledge and skills, giving the notion of "sea power" an even broader range. From the simple need to travel upon the waters have stemmed the fundamental challenges for naval architecture, navigation, meteorology, and hydrography. Drawing upon its experience within the British Empire, undisputed leading power in these matters during the eighteenth century, the young republic soon became an active participant and then partner with Britain over the two ensuing centuries. From such a practical maritime base, American science and technology expanded into other physical frontiers, exploring new waters and lands and studying in land, air, and space environments; into pure scientific research, especially in astronomy and oceanography; and into the applied areas of steam propulsion, modern ballistics, aeronautics, communications systems, medicine, and nuclear physics. All of these disciplines have merged with general American and world scientific developments, but one inescapable fact is that the sea and especially the U.S. Navy have been major catalysts.

To the naval architecture of sail, the United States contributed a great deal once it established an independent fishing fleet, Navy, and merchant marine. Pioneers in the fast fishing schooner from the early eighteenth century, the New England-centered Americans continued to excel in these two-masted workhorses throughout the 19th century. Such sleekness and speed also typified the early Navy's heavy 44-gun frigates, the revolutionary, record-setting 18.5-knot clipper ships of mid-century, and the racing yachts epitomized by the schooner yacht *America*'s victory in international competition in 1851. Even with the advent of steam, iron, and steel, the Americans continued to excel in sailing ship design, construction, and racing. The biggest wooden sailing ship ever built (reinforced with iron strips), was the four-masted, "jigger-rigged" ship and aptly-named *Great Republic* of 1853, built for trade with Australia. The big four-, five,- and six-masted schooners of the end of the century were topped off by the seven-masted *Thomas W. Lawson* of 1902, while the longest wooden sailing ship ever constructed, the 350-foot six-masted *Wyoming* of 1910, was built in this country. Finally, the America's Cup has been successfully defended down to the present, most dramatically by the fastest and biggest (Class J) yacht ever built, the 1937 *Ranger* designed by W. Starling Burgess and Olin J. Stephens.

Whereas they could be viewed as latecomers in the age of sail, the Americans were present at the beginning of steam propulsion, iron ship construction, and the general industrial revolution. Pennsylvania inventor Robert Fulton from 1793 worked on steam, canal, and submarine navigation. His efforts were crowned by the world's first practical commercial steam vessel, the *Clermont*, in 1807 and seven years later by the first steam warship, the *Demologos*. The first sail-steam schooner to cross the Atlantic, the *Savannah* in 1819, was American, and it was the American Collins ocean liners of the 1850s whose superior hull lines produced the best speed and motion through the water. Immigrant engineer John Ericsson designed the screw-propelled warship *Princeton* in the 1840s, the prototype of all modern steam vessels, and in the 1860s the celebrated *Monitor*, model for the first generation of the world's all steam-iron warships. Connecticut-born David Bushnell pioneered the submarine in 1775, and with the first real submarine of John P. Holland in 1900, the United States became a leader in this revolutionary field of naval design and construction. The nation also led in modern steel battleships, their armor upgraded by the American "harvey" process, and of aircraft carriers and many smaller warships. The epitome of the Navy's contribution to nautical propulsion systems has been the nuclear-powered ships and submarines of Admiral Hyman G. Rickover since the 1950s.

Just as immediate and profound as the practical effects of such progress in naval architecture and propulsion on the growth of the republic were concurrent developments in navigation. Massachusetts master mariner Nathaniel Bowditch built upon British work in nautical tables to publish in 1802 *The New American Practical Navigator*, which became the Navy's standard navigation text. To survey and chart America's long and expanding coastlines, five years later President Jefferson created the U.S. Coast Survey, which began (haltingly, to be sure) with the tides and currents, gradually worked inland to become the Coast and Geodetic Survey in 1887, and ultimately in the 1920s moved under and over the earth (seismology and air currents).

Without an established scientific community, the nation relied upon naval officers to promote this work during most of the nineteenth century. Consequently, Lieutenants Louis M. Goldsborough, Charles Wilkes, and James M. Gilliss developed the embryonic Naval Observatory with its navigational instruments and charts from its initial offices in 1830 and then official establishment in the 1840s, when Lieutenant Matthew Fontaine Maury became the first Superintendent and began his influential pioneer work in charting ocean and wind currents. In 1845 the Observatory initiated its first time signals to the American public, and twenty years later it first transmitted them nationwide via the new telegraph of Western Union. From the Observatory stemmed the Nautical Almanac Office (and its published annuals) under Lieutenant Charles Henry Davis in 1849 and the Hydrographic Office in 1866.

The crowning achievement of this practical use of the sea occurred in the fertile mind of Samuel F. B. Morse, the New York artist who in the 1830s invented the telegraph and conceived the idea of an underwater cable. In 1842 he laid such a line under New York Harbor, whereupon Congress funded his larger scheme the next year. Private capital enlarged his budget, leading to a temporary transatlantic telegraphic cable in 1858 and then, with important British help, a permanent cable in 1866. The impact of such mechanical communication under the sea on American—and world—progress can hardly be overstated.

To the science of ballistics and the general area of ordnance, the Navy has contributed importantly. From the "soda-bottle" muzzle-loading gun of Lieutenant John A. Dahlgren in the 1850s and 1860s to the powerful breech-loading 16-inch rifled guns of the last generation of battleships in the 1940s, the Naval Gun Factory and related agencies have supplied the armament for America's warships. When the range of naval firepower was extended with the invention of the airplane in America, the Navy lost no time in contributing to the growth of combat and civilian aviation and of ordnance. The same has been true of guided missilry and indeed of associated scientific and technological innovations such as radio, radar, loran, and computers, beginning with the Navy's IBM Naval Ordnance Research Computer in 1954. By the post-World War II era, the Navy's work in these fields had been absorbed into the larger arena of the new American technocracy.

Practical innovations in navigation were naturally supported by the American shipping industry and the Navy which needed such vital navigational data, freeing the young republic from utter dependence on the tables and charts of the Royal Greenwich Observatory and the Royal Navy. One must concede, however, that such an overriding concern for the pragmatic uses of astronomical data tended to thwart government support for pure scientific work in the nation. But it did lay the foundations for future developments along such lines. Also, such work served to initiate the country in the field of overseas exploration, notably into the Pacific and Antarctic waters before the Civil War and into the Arctic after it.

The advance from practical technology to pure scientific reserach and development proceeded rapidly from the late nineteenth century, with U.S. Navy playing a key role. The U.S. Naval Observatory and Nautical Almanac Office (finally rejoined administratively under the former in 1894) began the transition from nautical exploration to that of space. With the talented mathematician Admiral C. H. Davis as superintendent during the 1870s, the observatory installed the largest refracting telescope in the world with which the masses of the planets were calculated and the two tiny moons of Mars discovered. The observatory sponsored expeditions to observe solar eclipses and the transits of Venus, while Lieutenant A. A. Michelson first measured the speed of light at the U. S. Naval Academy. Admiral Davis' counterpart in the Almanac Office, the esteemed civilian astronomer Simon Newcomb, so improved the planetary tables for navigation over a 20-year period that in 1896 by international agreement they became the basis for the almanacs of every nation. And in 1904, the Naval Observatory adopted Guglielmo Marconi's wireless to initiate the worldwide transmission of its time signals via U.S. Navy radio stations, an important activity which has continued uninterrupted down to the present.

The final period of terrestrial exploration occurred with the union of the modern ship and the airplane, with Americans as leaders. Robert E. Peary began to explore the Arctic regions by sea and dog sled from the 1880s, pushing steadily northward until he reached the North Pole in 1909, after which he turned to aviation for the remainder of his life. The Navy wedded the flying machine to the boat hull to develop new seaplanes ("hydroaeroplanes") before World War I and utilized the seaman's knowledge of wind and sails to study aerodynamics. In 1919, the Navy's NC-4 flying boat became the first aircraft to fly the Atlantic. The navigational information for that flight was the work of a Naval Academy graduate, Lieutenant Commander Richard E. Byrd. He later applied the airplane to exploration by being the first person to fly over the North and South Poles, in 1926 and 1928 respectively. During the ensuing two and a half decades Admiral Byrd continued to lead the major American research and discovery efforts in the Antarctic. Simultaneously, it was the U.S. Navy which first published an air navigational almanac (1933) and which superintended the nation's program in rigid airships, a system which, however, failed in competition with heavier-than-air craft.

The zenith of sea power's impact on American and thus world scientific and technological progress arrived with the space age. The traditional stimulus of the sea in exploration has been easily transferred to what President John F. Kennedy in 1962 aptly termed "this new ocean." The means were also similar; the oceanic ship and the space ship are both closed eco-systems and administrative units with comparable logistical, medical, manning, and command situations. The same self-discipline and technical expertise for operating sea-borne-vessels pertains to space craft, so that the Navy immediately assumed a leading role alongside the Air Force and NASA in the American space program.

The celestial navigation techniques and tools pioneered by Captain P. V. H. Weems of the U.S. Navy were extended to the space environment, with navigational and communications satellites playing essential roles for ocean and terrestrial navigation and communications. The recovery area of American manned and unmanned space vehicles became the sea, with tracking ships linked to land stations and carrier forces retrieving reentered space craft. Finally, Navy-trained pilots provided a major element of the astronaut program. Lieutenant Commander Alan B. Shepard, Jr., became the first American to fly in space. In 1961, he manned the Mercury Freedom 7 capsule in sub-orbital flight before being recovered, by the carrier *Lake Champlain* (CVS-39). And in 1969, the Apollo program achieved ultimate success when former Navy pilot Neil Armstrong became the first human being to walk on the moon. The Navy has therefore played its part to the full in the space effort in the tradition of oceanic exploration. Just as the *Mayflower* brought the American colonists to our new world in 1620, it is their descendents who have sent the unmanned Viking lander to the surface of the planet Mars in this very month of July 1976.

History and the Sea

Without the presence of the sea, and the manifold challenges it has provided Americans, one simply cannot imagine how the republic and the race at large could ever have reached the level of scientific and technological brilliance that they have today. How this knowledge and sheer physical power will be ultimately utilized—for good or ill—of course depends upon human wisdom, and wisdom depends less on hard technology than on the human condition.

SOCIETY

At the helm, full sail,
Does mankind rush—
'Tis England, Olde and New—
The cargo hope—yea, promise.

The power of the sea has been a fundamental ingredient in the social life of the republic. More than any other single physical element, it has shaped the character of American society and has acted as a force of continuity throughout our history of change and uncertainty. It has broadened our horizons, intellectual and cultural, and has made us the envy and hope of the world. It has inspired our dreams and made us respect nature as no other force can. The sea has been our avenue to greatness, its intellectual power acting as the foundation of a folklore that will never die as long as the republic stands or is remembered.

From the first colonial settlements, America's cities have grown and prospered on the great watercourses of the continent. These cities have been the economic focus of our national life, the physical centers of our population, the meeting place of our businessmen and artists, the crossroads of American civilization. The merchants of New England and New York thus set the tenor for a capitalistic, cosmopolitan, bourgeois society, largely middle class and generally devoid of the elite aristocratic upper classes or suffering, hopeless peasant classes more typical of non-maritime nations. Even as the Northeast (and the agrarian South) declined in political importance with westward expansion, Wall Street lost none of its economic power, and New York City blossomed into the cultural capital of the nation. The great seaports thus remained both the jumping-off bases for American activity abroad and the front door for incoming immigrants, goods, and ideas from foreign lands.

Inevitably, the original cultural heritage of America—like its political and economic foundations—lay in its colonial maritime roots and has been centered in the Northeast from the earliest days. No better examples can be found than in two great American families, Adams and Roosevelt. Both are inextricably linked to the sea, its art and history, to business and politics, and to the American dream, all of which their members have championed. The original Englishman Henry Adams came to Braintree, Massachusetts, in 1636 and ultimately fostered, among many luminaries, Presidents John Adams (1797–1801) and John Quincy Adams (1825–1829), statesman Charles Francis Adams, historians Charles Francis Adams, II, Henry Adams, and Brooks Adams, and banker-yachtsman Charles Francis Adams, III, who was also Secretary of the Navy (1929–1933). The Roosevelts can be traced to a Franco-Dutch settler of the Plymouth Colony in 1621 and a Dutchman of New Amsterdam (New York) about 1649. From that beginning ultimately emanated Presidents Theodore Roosevelt (1901–1909) and Franklin D. Roosevelt (1933–1945). Both were Assistant Secretaries of the Navy as well (1897–1898 and 1913–1920 respectively), as were Theodore Roosevelt, Jr. (1921–1924) and Henry L. Roosevelt (1933–1936). Like the other Massachusetts and New York businessmen and leaders around them, the Adamses and Roosevelts reinvested their material profits into the arts and humanitarian purposes in general. They were patrons of the spiritual components to material prosperity.

The rich literary heritage of the New England community arose from the early nineteenth-century authors and poets who wrote near and/or about the sea. America's first true philosophical movement of Transcendentalism centered here in the works of Ralph Waldo Emerson and Henry David Thoreau, while Herman Melville in his *Moby-Dick* tried to fathom the mysteries of the sea and men against it. Ordinary seaman Richard Henry Dana did likewise in *Two Years before the Mast*, and Henry Wadsworth Longfellow used the ship as a metaphor of the republic when in 1849 he wrote, "Sail on, O Ship of State. . . ." The likes of novelists James Fenimore Cooper and

Nathaniel Hawthorne and poets Walt Whitman and John Greenleaf Whittier helped give America its literary character, and young Oliver Wendell Holmes, Sr. helped save the uss *Constitution* from being broken up with his moving poem of 1830, "Old Ironsides."

These traditions have survived undiminished throughout the twentieth century. In the 1920s, Melville's lost novel *Billy Budd* received critical acclaim. Harvard graduate Charles B. Nordhoff, whose grandfather had sailed in Yankee windjammers, moved to Tahiti with James Norman Hall where in the 1930s they produced the epic trilogy *Mutiny on the Bounty, Men Against the Sea,* and *Pitcairn's Island.* The most successful recent fictional works have been the penetrating novels *The Caine Mutiny* and *Winds of War* by Herman Wouk. And one must not forget the unsurpassed historical prose of Harvard historians Samuel Eliot Morison of Massachusetts and Robert Greenhalgh Albion of Maine. Their histories of Americans and Europeans at sea have flowed continuously from their pens over the past half-century.

As for painting, generations of artists, following the example of Winslow Homer, have captured the power and majesty of the sea from the vantage points of the Maine and Massachusetts coasts. One such individual, John Marin, in a reflective moment in 1938, perceived the unique advantages of the republic's proximity to the sea when he observed, "Isn't it funny that Dictators *never never never* live by the sea?" While a storm pounded the shore, he considered the fascist rulers abroad: "Don't lock the tyrants up—make them look at this great sea manifestation taking place." The awesomeness of the ocean environment—portrayed so vividly, for example, in the canvases of C. G. Evers—has been understood by such artists as a special force in the shaping of unique American social qualities. So the quiet fishing village scene of downeast Maine tells as much of the people as does a mural showing a square-rigged ship under full sail in a gale. It captures the very quality of individualism that has been the basis of American democracy, of this Columbia, the Gem of the Ocean.

The severe demands on the merchant and warship sailor to survive have been common to all seafaring nations, but it has been the great liberal maritime democracies of Britain and the United States which have pioneered in humanitarian reforms ashore and afloat. The new republic has led in many respects, however, in keeping with its strong libertarian ideals. The unpopular British practice of impressing common seamen by force was abandoned largely at American insistence in 1815. The British habit of the grog ration outraged American temperance reformers and Congressmen who thus outlawed it in the U.S. Navy in 1862, while Secretary of the Navy Josephus Daniels abolished all alcohol on board ships in 1914. Needless to say, such a purification scheme was and still is regarded with much skepticism by thirsty swabs, but it was all a part of the very necessary effort to clean up the harsh conditions of shipboard life, to eliminate alcoholism, and thus to enhance performance of duty. Besides, enlisted men's and officers' clubs ashore, liberally rationed medicinal brandy, and privately-kept liquor under personal lock and key at sea (and ingenious concoctions of "torpedo juice" in World War II) have compensated for the official "dry" policy. The need for improved diet, medicine, and general health at sea, along with the social, educational, and religious requirements of the individual sailor, have kept the Navy in the important business of providing essential services to the average sailor, most recently in the area of drug control and rehabilitation.

Life at sea and along the urban waterfronts was still rough-and-ready even as late as the turn of the present century. This writer well remembers his late grandfather, Gilbert P. "Curly" Reynolds, fireman first class and coal passer in the armored cruiser *Maryland* (1912), tattooed eagle emblazoned across his chest, telling how he once flattened a junior officer one day after the latter had kicked him as he scrubbed the deck. Grandpa spent some time in the putrid brig until the truth became known and the officer replaced him there. He and his mates of that era worshiped a man like Admiral Robley D. "Fighting Bob" Evans, whose son Captain Franck Taylor Evans was a chip off the proverbial old block. One day about 1913, after a shore leave at the notorious "Barbary Coast" of San Francisco in which a number of his men had been beaten up and robbed by the proprietors, their bouncers, and ladies, this Evans led a shore party which tore up the place and henceforth gained his sailors the respect they deserved while ashore. My other grandpappy, Charles C. Clark, still vividly recalls an occasion in Seattle when bluejackets from the Great White

Fleet rallied to their threatened comrades in the red light district in like manner.

A similar problem which appeared in 1917 had a profound impact on American music. The port was New Orleans, already infamous for its "Storyville" red light district, but under wartime conditions a menace to young seamen on liberty. The musical entertainment at the "sporting houses" was provided by black musicians playing the new type of jazz music which was in fact becoming America's only original contribution to the arts. The Navy Department closed down the district as a wartime measure, forcing the jazzmen to seek new employment up the Mississippi River, on the river boats, at Memphis, Kansas City and Chicago, thence to New York. Thus did jazz begin to sweep the nation and the world.

Eventually, the music matured into "swing" and found its way back to the Navy, as each major warship during World War II formed her own swing band. In the tradition of U.S. Marine bandmaster of the 1880s John Philip Sousa and his stirring marches, these orchestras were terrific morale builders, as were "name" touring units like that of Navy bandmaster Artie Shaw which toured the South Pacific or Saxie Dowell's group as it played on board the bomb-stricken carrier *Franklin* (CV-13) off the Japanese coast. The crowning glory occurred in England in 1944 in a "battle of the bands" in which Sam Donahue's U.S. Navy band thoroughly outplayed Glenn Miller's Army Air Forces orchestra.

Progress in twentieth-century photography, still and motion, greatly enhanced the image of the sea in the public's mind, making legendary heroes at once even more larger than life and more personal, especially via television. From the novels noted above and the Broadway stage came Hollywood scripts, as well as original screenplays. In the 1930s, the sailing ship era received much attention through the swashbuckling performances of such characters as Errol Flynn, and the prewar Navy received a romantic image through dramas like "Helldivers" and even musicals like "Follow the Fleet."

But World War II marked the pinnacle of the naval film, between the documentary stills of Captain Edward Steichen, documentary movies like Commander John Ford's filming of part of the Battle of Midway, and Lieutenant Dwight Long's story of the new carrier *Yorktown* (CV-10), "The Fighting Lady." Many fictional/semi-documentary movies starring such actors as John Wayne, Dana Andrews, and Tyrone Power captivated American audiences during and long after the war. So motion pictures from "A Wing and a Prayer" to "Sands of Iwo Jima" to "In Harm's Way" not only entertained but educated the public and even proved to be effective recruiting devices. Cinematized Pearl Harbor has kept Americans sensitive to sneak attack in dozens of films, culminating in the epic "Tora! Tora! Tora!" Ship types received their due from PT boats ("They Were Expendable") to carriers ("Task Force") and especially submarines, through the eyes of the attacker and the victim ("Destination Tokyo," "Action in the North Atlantic," and "The Enemy Below").

And since the war, the Navy has provided an even broader stage for portraying on film the human drama as embodied in the republic's actions on the waters. The documentary television series "Victory at Sea" even utilized a brilliant musical score by Richard Rodgers. The Korean War found expression in "The Bridges of Toko-ri," based on James Michener's book. He had earlier written the story behind the successful Broadway musical "South Pacific," the score by Rodgers and Oscar Hammerstein. Potential nuclear war found naval settings in "The Bedford Incident" and "On the Beach," Hollywood's version of Armageddon. And on a more optimistic note, "Mister Roberts" and "Kiss Them For Me" (based on Frederic Wakeman's novel *Shore Leave*) captured the great good humor and human pathos of the American sea fighter in war.

Between the old and young salt represented by John Paul Jones, Captain Ahab, Bull Halsey, and one Ensign Pulver emerges the composite of a naval character unique in the annals of history, a uniqueness which can best be understood by the term "American." Until the British naval bombardment of Falmouth (Portland), Maine in 1775, Americans had so revered their naval heritage that the Washington family of Virginia could name its plantation in honor of a British admiral, Edward Vernon. Post-Revolutionary naval heroes were similarly honored, but they were never given political authority or prestige, nor did they want it. The Secretary of the Navy held absolute control over the Navy down to World War II, while the rank of admiral—considered aristocratic and un-American—was not even created until the 1860s.

Indeed, the nation's sailors have stayed generally out of the political arena—as all navies do—preferring instead to practice their art far away, at sea. The only time a flag officer was even mentioned as possible Presidential material—George Dewey in 1900—was an aberration which lasted but a few weeks without serious impact. In fact, just as the United States at the turn of the century created a major standing battle fleet officered by products of the highly professional U.S. Naval Academy (from 1845) and Naval War College (from 1884) and who maintained their closest dialogues in their own private U.S. Naval Institute (from 1873), the Navy began to foster nonprofessional sources of officer and enlisted manpower: the Naval Militia (1888–1918), the U.S. Naval Reserve (from 1915), the Naval ROTC (from the 1920s) and Naval Aviation Cadet programs (from the 1930s). With the coming of World War II, the Navy finally expanded into an institution representative of a true cross section of the republic—typified by the fact that the last four Presidents, John Kennedy of Massachusetts, Lyndon Johnson of Texas, Richard Nixon of California, and Gerald Ford of Michigan, were wartime Naval Reserve officers.

The U.S. Navy has become a major component of the contemporary American consciousness, but at the same time it has maintained its traditional political aloofness. Unlike Army and Air Force officers who deal with large administrative organizations and masses of personnel and who thus develop political savvy and clienteles which can project them into high political office, regular Navy personnel have stayed oriented to shipboard activities, out of touch with landlubbing politics. Retired admirals may utilize their logistical expertise in corporations, but few have had sufficient talents to achieve elective office. They are by nature and preference seamen first and foremost, for which the republic can be grateful.

By the same token, however, naval leaders often suffer in political battles with more politically-seasoned civilians and generals. Some of the more well-known examples are Admiral James O. Richardson's dismissal by FDR when the former tried to remove the U.S. Fleet from Pearl Harbor just prior to World War II, Admiral Ernest J. King's repeated failures to reorganize the Navy Department during the war, the "revolt of the admirals" against postwar unification, and Admiral George W. Anderson's relief over the TFX controversy. Notable and rare exceptions have been Admiral William A. Moffett's successful crusade for naval aviation in the halls of Congress during the 1920s and 1930s, Admiral William D. Leahy's role as chief of staff to the President during the 1940s, and Admiral Hyman Rickover's dogged but successful campaign for nuclear powered warships throughout the government and industry over the past thirty years. The Navy's success in Congress has depended almost exclusively instead on civilian leaders like Carl Vinson who have understood and appreciated the Navy's missions.

The subtlety of sea power has been the major stumbling block to the public's developing a full awareness of its importance. The vision of a naval battle is real enough, but Congressmen and Presidents have traditionally had precious little sympathy for the financial needs of the fleet and the merchant marine, vital as these are. Even when a major naval philosopher emerged in the person of Captain Alfred T. Mahan in the 1890s, his ideas were more oversimplified and uncritically canonized than really understood—inside the Navy as well as outside. Naval historians in general have been read for their action narratives, not for their philosophical insights. And even in the wake of naval victories in all of the nation's major wars, immediate, rapid, and dangerous postwar naval demobilization has inevitably followed.

The republic appreciates the power of the sea only superficially, but that influence endures more importantly now than ever in the past. We underestimate or ignore it only at our peril.

The final measure of any nation's power upon the sea is its ultimate impact on the race at large. The original seafaring nation or thalassocracy was fifth-century B.C. Athens, whose democracy, empire, economy, and culture not only dazzled the ancient world but laid the foundations for all Western civilization. Though the state of Athens was eventually destroyed politically by jealous rivals and through internal decay, Athenian culture conquered Alexander the Great and his successors (fourth to second centuries B.C.) and the Roman Republic and Empire (second century B.C. to fourth century A.D.) and provided the seeds of the Italian Renaissance (fourteenth to sixteenth centuries). This latter event centered on maritime Venice and neighboring Florence and to a lesser extent on seafaring Genoa and Pisa—all republics which rescued Europe from the stultifying dark ages and from Oriental enemies, propelling Western civilization into the modern age. The torch of enlightenment passed by sea to the republican Netherlands and democratic England (seventeenth and eighteenth centuries), the latter spreading this Western culture across the globe via its ships.

The United States—a democratic republic—has gradually replaced Great Britain as the defender of Western ideals and culture, the model of aspiring nations in political freedom and economic prosperity. On three occasions, World Wars I and II and the Cold War, the republic rescued Western Europe from the threat of authoritarian overthrow and cleared the Pacific of similar dangers. But victory was only the beginning, for conquered Germany and Japan were transformed into giant models of the American colossus, with Soviet Russia and Communist China and every developing small nation unabashedly seeking American aid or imitating American techniques in the hope of achieving those material advantages which Americans hold dear. They thus take to the sea in search of wealth. If they succeed fully, however, they will prosper in many more ways than simple treasure. For the American Republic is also rich in the common human aspirations held sacred by all freedom-loving people since the beginning.

> *O Sea! Eternal Force!*
> *Sing, Voyagers;*
> *We're upward bound,*
> *For God has smiled.*

The republic is not a utopia. Though motivated by the noblest of human ideals—life, liberty and the pursuit of a happy life through material well-being—two centuries ago, Americans have shown themselves to be heir to the same frailties that plague the race at large. We have made mistakes, abused power, often fallen short of our ideals. Even when our intentions have been well-meant, we have often stumbled, out of ignorance or inexperience. And yet, painfully and persistently, we have striven to overcome our deficiencies, to correct our mistakes and even to learn anew. Perhaps this challenge is the source of the drive and the excitement that keep us active, growing and prosperous.

And yet, there is something more than mere challenge, something unique, subtle and unspoken, that continues to shape and propel us, something that must account for our success where others have failed. In the view of this historian, it has been the sea—its mere presence, its riches, its horizons, and its awesomeness. Other challenges and other factors have influenced our history, to be sure, and often they have eclipsed our awareness of the waters. And yet, we always return to the sea—that vital force in the making of America, its importance no less now than 200 years ago or at any time in our past.

5

RECONSIDERING AMERICAN
STRATEGIC HISTORY AND DOCTRINES

The Mahan-influenced views of American military and naval historical writings and interpretations became apparent to me as I reshaped my appreciation of the United States as a thalassocracy, as shown in chapter 4. Specifically, I realized that most historians had either accepted his ideas uncritically or had tended to reject his arguments and thus to regard America's wars and defense policies from the standpoint of the U.S. Army and land warfare. I therefore wrote the present chapter as a heavily annotated scholarly reinterpretation of America's military past in terms of "strategic" history, incorporating political, economic, diplomatic, army, navy, and air force doctrines.

The chapter, originally entitled "American Strategic History and Doctrines: A Reconsideration," appeared as an essay in the December 1975 issue of *Military Affairs*, the premier journal of American military historians published by the American Military Institute. It won the AMI's Moncado Prize.

Between C. Vann Woodward's appeal in 1959 for a general reinterpretation of American history in light of new military realities and Correlli Barnett's relabeling in 1969 of "the study of organized conflict in human relations" as "strategic history," America's military past began to take on new definition and meaning.[1] No longer "unhappily adrift," as Walter Millis lamented in 1961, American military history has been profiting from new historical and theoretical perspectives made possible by the momentous events of the 1960s and 1970s, evolving toward a new maturity worthy of Barnett's label.[2]

Specifically, strategic history allows a general reconsideration of America's (or any country's) military doctrines by seeking to reveal patterns of continuity in the strategic actions of types of nations and as a consequence to understand the behavior of each particular nation. It, therefore, makes possible historical hypotheses based not so much on the more traditional episodic war history, institutional administrative history, or isolated national history as on fresh sweeping overviews of the past. A half-century ago, Hans Delbrück called for such analyses to throw light "on the interconnections in universal history," a subject more in vogue then than now; nevertheless, the new strategic history offers the highest possible interaction between past organized conflict and general history and—in that context—a fuller specific understanding of United States history.[3]

For meaningful hypotheses, strategic history must draw upon three general kinds of sources: traditional, more conventional military histories;[4] interrelated sister disciplines and techniques such as civil-military political studies, sociology, psychology, economic theory, and quantification;[5] and the works of key strategic theorists and non-military scholars past and present who are concerned with universal issues.[6] As one example which draws on all three, Theodore Ropp has borrowed from historian of science Thomas Kuhn certain methodological techniques to examine the sweep of the past two centuries of actual armed conflict in terms of "modern military paradigms," which in turn suggests strong overtones of generational theory and history.[7] The strategic historian is therefore less concerned with stated policies, professed needs, official doctrines, or beliefs of particular nations than with their *actual use* of armed forces in wartime *and* peacetime.

A complete and realistic utilization of strategic history has not been possible in the United States until recently for several reasons.

History and the Sea

American historians, traditional as well as revisionist, like the people whose history they have written, have tended to be internal-oriented, rather than regarding their nation's past in terms of world or universal history. Such an isolationistic or unilateralistic bias has been understandable given the strategic "free security" of pre-1941 America. But this country's more recent world leadership—however reluctant—means that the real strategic questions of the past and indeed the future can no longer be begged. Viewed from an internal perspective, for instance, the United States has shown its democratic preference for a peacetime militia—or wartime citizen-based army and a small navy—over a large standing professional army and battle fleet.[8]

In strategic terms, however, this preference for a low military profile has been realized as much—if not more or so—by strategic realities as by popular attitudes, as the present essay will show. Both factors have led to a rich Army-centered military history and defense policy throughout most of American history. The result has been a largely discontinuous (and therefore artificial) treatment of American strategic history both by the large and productive U.S. Army's Center of Military History—geared more to "a narrow operational and tactical context" than to "considering broad integrative history"[9] and by otherwise excellent Army-oriented civilian scholars who base their work on the short-sighted and short-term assumption of the "episodic or intermittent character of warfare" and its "brevity, intensity, and relative infrequence."[10] Consequently, for example, the two just-quoted military scholars, Peter Paret and John Shy, could assume, incorrectly, in 1962 that a "new dimension was added to American military policy" in 1961–1962, namely "limited conventional war."[11]

Strategic history shows, in fact, that the American military experience—broadly and just the Army's—has been more continuous than episodic and that the major strategic doctrinal continuity has been one of waging limited conventional warfare. The "episodes" indeed have been the exceptions to the norm: the Civil War, World Wars I and II, and the Russo-American thermonuclear Cold War. The public's and historians' preoccupation with the more direct and glamorous total wars not only distorts the actual continuum of American strategic doctrines but leads to a historiographical and theoretical evolutionary dead-end whereby Army-oriented scholars like Russell

F. Weigley conclude, again incorrectly, that "the history of usable combat may be at last reaching its end."[12]

American strategic history has also suffered from a serious and closely related shortcoming until quite recently, namely the subordination of naval doctrinal history to the Army's (and Air Force's). The subtle and less obvious applications of naval power, along with Americans' distrust of the political authoritarianism reminiscent of the Royal Navy's admirals,[13] are part of the reason. Another is the tendency of the U.S. Navy's small Office of Naval History and of naval-oriented civilian scholars—like their Army counterparts—to stress an episodic approach and even to fall into the same total war evolutionary trap noted above. And Army-centered historians have tended merely to borrow heavily from such conventional naval history to reinforce their own findings.

Not until 1956 did any survey of American military doctrines include a balanced Army, Air Force, *and* Navy (including the Marine Corps) treatment: Walter Millis' *Arms and Men*. But Millis acknowledged guidelines set by Harold and Margaret Sprout in 1939; their *Rise of American Naval Power* was "something of a turning buoy to a new course" which sought to correct Americans' episodic views of war and military institutions by regarding such phenomena as "continuous factors within the fabric of our society."[14] But Millis then, and the Sprouts ten years later, recognized one peculiar failing of the latter's key work and of the works of most American naval historians, namely, an uncritical obedience to the doctrinal and propagandistic theories laid down by Alfred Thayer Mahan in 1890: the belief that the United States had erred in not creating a blue-water battle fleet before the twentieth century.[15] The Sprouts, however, unknowingly anticipated the coming "new order of sea power" by suggesting in 1940 that "the American people, with their immeasurably greater resources, might conceivably succeed in establishing a virtually global dominance" at sea.[16] With the realization of such a "new order" after World War II, the way was clear for Millis' book and indeed for a new synthesis of American strategic history based on global and naval considerations as well as on those of armies and air forces.

New hypotheses—the intent of the present essay—regarding America's military past are now possible within the context of strategic history, not only because of the timeliness and evolving scholarship discussed above but also because of a great wealth of relevant monographic work, especially in doctoral and masters theses, and in contemporary strategic theory in the United States and Great Britain over the past fifteen years. This writer's first hypothesis, briefly stated, is one of definition and categorization: strategically, two general types of great powers have existed throughout history, *maritime* and *continental* states, with the *small* nation being in a separate class by virtue of its strategic dependence on the protection of an allied great power. Because of national political, commercial, and actual defensive needs, the maritime state has followed a maritime strategy, its navy as the senior service, its army secondary, and an allied continental army providing the major hitting force against large enemy armies on continental land masses. Conversely, in the continental state, the army has been senior and the navy subordinate. The second hypothesis is one of application: strategically, the United States has been a *maritime* nation throughout its long history—perhaps not consciously, overtly, or unilaterally, but in action and effect most assuredly so, beginning as a small power but growing into great power status. The proof of these hypotheses lies in the vast array of convincing scholarship now available to test them, though only the latter thesis is the concern of the present essay.[17]

The overriding strategic fact throughout American history down to World War II—one usually ignored or passed off superficially—is that the first line of national defense lay not in America's Army or Navy but rather in the British Navy, creating the condition which Woodward has called America's "free security":

> The costly navy that policed and defended the Atlantic was manned and paid for by British subjects for more than a century [1815–1941], while Americans enjoyed the added security afforded without added cost to themselves . . ., a security so complete and so free that [the United States] was able virtually to do without an army and for the greater part of the period without a navy as well.[18]

Indeed, considering that the American people were an integral part of the British political empire throughout the century and a half of the colonial period, and even fought a mutual war against France in 1798–1800 in the midst of an otherwise stormy Anglo-American diplomatic period, the strategic continuity and durability are the more understandable.

American strategic history must be understood in these terms, not only to make America's military past comprehensible but to understand the current era, "the new order of sea power," wherein global maritime leadership has passed from Britain to the United States. Strategically, this order is the same as during the nineteenth century, only the label is now *Pax Americana* instead of *Pax Britannica*.[19] The theory and practice of Anglo-American maritime strategic doctrine therefore must be studied both in a broadly theoretical perspective from the standpoint of the British position and in detail from the actual experiences of American soldiers and sailors over the past three-plus centuries.

The theory and often uneven practice of maritime strategy shared by the British and American people has been based upon the need and desire for profits through overseas trade. Whether under a closed mercantilistic system as in the eighteenth century or the open free trade typical of the nineteenth, Anglo-American ocean-going merchantmen have cruised the sea lanes of the world under the general protective strategic umbrella of naval power. When threatened or attacked during the early years by pirates, privateers, or enemy cruisers, the merchantmen received protection from especially established regular naval squadrons whose function was to police such troubled waters and to operate from nearby bases which had been taken or leased to serve them. Except in the waters around Europe, where a British fleet of capital ships maintained a general deterrent presence against any great power challenge to the maritime order, such overseas squadrons were comprised of lesser vessels more suited to inshore, pacification-type operations—frigates or cruisers and gunboats. The British empire included overseas territorial colonies, while the American empire was largely continental until the 1890s when briefly it, too, embraced overseas colonies. Both countries, however, intervened ashore to suppress slaving activities, pacify piratic native peoples or insurgents, and stabilize—politically and economically—regions where the trade was important to both capitalistic nations.

The British and American ground forces involved in such operations were usually professional frontier garrisons as in India, Africa, or the American West or naval brigades of professional sailors, marines, and soldiers in coastal regions. Imperial or overseas policing, pacification operations, and limited inshore warfare became the norm, continuous over the decades and centuries, right down to the present. The world wars and larger localized wars (Crimean, American Civil, Boer) required military expansion via a *levée en masse* of nonprofessional militia-National Guard-reserve call-ups, volunteers and/or conscripts, but neither nation operated on the European continent without a major continental allied army bearing most of the responsibility there. By thus examining these actual roles and missions of the armed services of the American people in light of the British example, the strategic historian discovers remarkable similarities in their strategic doctrines and tactical practices.[20]

American strategic history can be divided roughly into three periods: the small-power period of the "old order," the seventeenth, eighteenth, and nineteenth centuries, during which the Americans fit within the general British strategic sphere as colonists or as a small nation, despite the wartime intervals of 1775–1783 and 1812–1814; the period of Anglo-American cooperation, however erratic, 1900–1947; and the "new order" of American leadership under the *Pax Americana*, since 1947. Through it all, Anglo-American command of the sea has guaranteed America's strategic insularity, which has made standing naval forces absolutely necessary and army forces of secondary though still vital importance.[21]

The American colonial experience established important doctrinal habits within the context of British strategic policy which conditioned American strategic policies throughout the "small-power" period. The Royal Navy gave general protection to the American colonies, fighting the French (and often their Spanish allies), until eliminating them from North America by 1763. What little actual naval activity the colonists contributed was in the form of privateers and armed merchantmen for *guerre de course* (commerce raiding) as very minor auxiliaries of the Royal Navy.[22] By the same token, colonial militias reflected an earlier English militia system and evolved into a civil-military institution that became the basis of the U.S. Army until 1815—with influences lingering much longer. During the seventeenth century, defense against Indians provided the major military impetus, but the Dutch menace led to the posting of "independent companies" of British regulars from the 1660s to complement the militia. In the subsequent wars against France, the colonial militia and independent companies assisted the Royal Navy and British armies in providing imperial defenses, with the militia of Massachusetts being especially vigorous in Anglo-American operations to Nova Scotia as part of that forward defense during the 1740s and 1750s. Militia units were also active in internal "civil" actions, in Virginia during Bacon's Rebellion (1675–1676) and in South Carolina helping to control the slaves.[23]

After the Seven Years War (1756–1763)—more aptly named in America for the French and Indian enemies—the British withdrew the independent companies and tried to weaken the militia by turning over the North American defenses and internal policing to the Navy and regular Army battalions, a policy that helped to trigger the Revolution. In military doctrine, however, the forerunners of the Continental and U.S. Armies had established two primary functions: coast defense against European enemies and frontier policing against the same foes and Indians allied to them or acting independently. The former mission was episodic as part of Britain's major wars, while the latter was continuous, unglamorous, and bloody, involving the militia—which often adopted Indian techniques in irregular warfare—in what might be relabeled as defensive imperial pacification operations.

The first forty years of independent American strategic practices, 1775–1815, were but a continuation of the colonial doctrines, both services concentrating in general on coast defense (including the lakes), and in particular the Navy on commerce protection and raiding and the Army on frontier pacification. During the Revolutionary War, the French Navy replaced the British Navy to achieve command of the sea just long enough to guarantee American independence, with the small Continental Navy in a supporting *guerre de course* role.[24] But the Continental and United States Navy, the latter created officially in 1798, followed British naval practices, habits, and shipbuilding designs during and after the Revolution, even allowing the Royal Navy to convoy American merchantmen during the Quasi-War with France (1798–1800), and followed British naval leadership in the West Indies during and after that struggle. Jefferson's gunboats and frigates emphasized coast defense and commerce warfare against the Barbary pirates, but always (save for the War of 1812) within the general context of overall British strategic protection.[25]

As for ground forces, in the Revolution, the Continental Army (reinforced by French regulars) formed the nucleus of the fighting forces on land to fight pitched battles, but—significantly—it often joined the Indian-blooded irregular forces of the interior to harry the enemy in guerrilla-type fighting. Indeed, says Don Higginbotham, ". . . even though the patriots never turned to bush fighting as their principal means of resistance . . ., [in] a prolonged conflict . . . Americans would probably have turned increasingly to guerrilla warfare. . . ."[26] After the Revolution, the Indian wars continued, while the struggling young Republic also required troops to maintain domestic stability during two "rebellions"—Shays' and the Whiskey (1786–1787, 1794). But it was the Indian struggles, from Fallen Timbers (1794) to Tippecanoe (1811), which provided the catalyst and real experience for shaping the professional U.S. Army as a more sound alternative to the less disciplined frontier militiaman.[27] Without an allied French expeditionary force or supporting French fleet in American waters, but with a larger and more effective professional Army and Navy, the Americans fought the War of 1812 on the same ground and waters of the Revolution in a decidedly limited conflict, which only served to reinforce the established American strategic doctrines of coast defense, merchant warfare, and frontier pacification.[28]

"For a century after the Peace of Ghent," observes Harry Coles, "the Royal Navy was the main shield of the American Republic against the distresses of Europe."[29] The Monroe Doctrine rested on that strategic foundation, basically uninterrupted between 1815 and 1940 and augmented by an upgraded system of coastal fortifications and several overseas cruiser squadrons that operated in more-or-less tacit cooperation with the Royal Navy in the protection of British and

American traders under the *Pax Britannica*. Despite several diplomatic crises, Britain defused potential armed confrontations by accepting what became the current contiguous borders of the United States and by virtually disarming the Canadian frontier.[30]

With its strategic insularity thus preserved, the United States utilized its professional armed forces for their other traditional role of frontier pacification. The Army, in spite of its considerable improvement in technical, European-oriented training at West Point after 1817 under Sylvanus Thayer and Dennis Hart Mahan,[31] waged an almost continuous war against the Indians (and outlaws) in protection of American settlers, from the First Seminole War (1817–1818) to the tragedy of Wounded Knee (1890). The frontier forts acted as virtual colonial outposts of the American empire, connected to the states by riverborne logistics. In addition, the Army and Navy cooperated for inshore operations in the Second Seminole War (1835–1842) and the Mexican War (1846–1848) in the first "overseas" aspects of empire building—the pacification of the Florida Peninsula and the conquest of the upper California coast. In all these operations, volunteers and militia merely augmented the professional services in their traditional roles in the defeat of Mexico.[32]

The one towering exception to this century long continuum of strategic doctrines was the Civil War experience, a conflict, in Michael Howard's view, "of even greater significance for world history than those of Napoleon and von Moltke" because of the subtle factors of "blockade, conscription, war production, attrition of manpower and morale . . . which were beginning to count. . . ."[33] These larger and more dramatic aspects of the war, therefore, naturally deserve the attention they have received, but of great importance to a fuller understanding of the events of 1861–1865 are the continued doctrines of the prewar years alongside the episodic demands of this first modern war. The U.S. Navy seized command of the sea almost by default, turning to blockade Southern commerce, while the Confederacy adopted the traditional American doctrinal methods of coastal defense and *guerre de course*. Otherwise, the Union Navy provided logistical mobility to armies which, significantly, were all named after the rivers in their operating areas. The volunteers in America's only true strategically continental war provided most of the manpower for both sides to fight campaigns on the relative scale of Napoleon's wars and World Wars I and II, though much of the regular U.S. Army and many volunteers remained on the frontier.[34]

The Indian wars reached a new peak during these years, while Confederate guerrillas operated with initial effectiveness in the Mississippi Valley and elsewhere—so much so that the rebel government could consider a prolonged guerrilla war as a viable alternative to surrender in 1865.[35] Professional and volunteer Army and Navy forces alike waged pacification operations annually against the Sioux and other Plains Indians, continually against rebel guerrillas in northwest Tennessee and throughout Missouri, and occasionally in Virginia against "Mosby's Confederacy." Especially telling were—in modern parlance—the early-day waterborne "search and destroy" operations of the Army's specialized Mississippi Marine Brigade along that river.[36] And, at war's end, the Navy returned to its distant stations alongside the British, and the U.S. Army had to police the occupied defeated states, fight the Indians, help—along with the National Guard units—quell labor riots, and quietly pursue the larger doctrinal and professional questions of modern war.[37]

The period of Anglo-American strategic cooperation actually began during the 1890s when the United States embarked on its one foray into overseas territorial imperialism, with Britain—partly because of changing power relationships in Europe and Asia—choosing to regard America's overseas thrust not as a danger but rather as complementing the defenses of the British Empire. Following the American victory over Spain in the decidedly limited maritime conflict of 1898, determined largely by naval blockade, naval battles, and amphibious operations, anti-British rhetoric in the United States virtually disappeared. And the U.S Navy developed—for the first time in its history—a major battle fleet of capital ships with its own blue-water capability for commanding the sea. The British navy still provided America's principal strategic shield in the North Atlantic, while the U.S. Navy replaced the Royal Navy as policeman of the Caribbean. Both nations regarded Germany as the main potential enemy, though each did so unilaterally and not within any formal alliance.

By the same token, neither Britain nor the United States could afford the sheer cost of providing substantial naval forces in the Pacific, despite the American acquisition of the Philippine, Wake, Guam, and Hawaiian islands. Japan's naval victories over China in 1895 and Russia in 1905 made that nation the major Pacific power, which Britain joined in formal alliance in 1902 and which the United States accommodated by diplomatic concessions and occasional naval demonstrations such as the voyage of the Great White Fleet in 1908. With its own imperial territories in the Pacific and Caribbean, and following Alfred Thayer Mahan's advice, the United States created a maritime strategic stance closely akin to Britain's. Americans could agree, from 1900 to the present, with Mahan's statement in 1906 that, "taking the constitution of the British Empire and the trade interests of the British islands, the United States has certainty of a very high order that the British Empire will stand substantially on the same line of world policy as ourselves."[38]

To argue, however, that the United States government between 1898 and 1917 consciously developed a systematic, centralized, and logical overall maritime strategic doctrine would be futile. Nevertheless, the sum total of the actual activities of America's armed forces from the Battle of Manila Bay to the expedition against Pancho Villa reveal such a strategy in actual practice. The Navy, particularly under the Presidential leadership of Theodore Roosevelt, became the senior service, charged with commanding the Caribbean, policing the waters of America's new colonies and the Western Hemisphere, and using battleships to demonstrate unilateral American power in Europe and the Pacific against Germany and Japan. Partial administrative reform, inspired somewhat by the British example, led to the advisory General Board in 1900 and the Office of the Chief of Naval Operations in 1915 and to the creation of the Atlantic, Pacific, and Asiatic fleets in 1906–1907. To provide the necessary amphibious forces during the interventions in the Caribbean, as well as naval base defensive garrisons, the Navy enlarged the role of its Marine Corps from mere shipboard policing into an integral component of American maritime strategy.[39]

That the U.S. Army did not assume more of an active role in such overseas interventions after the war with Spain has been explained by its traditional preoccupation with the coast defense of the United States and a concurrent lack of interest in the preparation for such expeditions. The Army, influenced by European trends, reorganized under Secretary of War Elihu Root in 1903 with a General Staff and Chief of Staff, but in practice the Army officers who began their careers during the last Indian wars spent them largely in similar occupation and pacification operations in Puerto Rico and Cuba (1898–1902; the "Army of Cuban Pacification," 1906–1909); the Philippines (Insurrection, 1899–1902; anti-Moro operations and the Philippine Constabulatory, 1901–1913); and Mexico (1914, 1916–1917, 1919), all of which can be closely related to the doctrinal defense of American territory.[40]

By the outbreak of World War I in Europe in 1914, the United States had achieved great power status, militarily as well as economically. In theory, the basis of American military strength lay in the Army, specifically the citizen-armies of militiamen and volunteers which had won every war, but most dramatically the great struggle of 1861–1865, a tradition which the annual encampments of the Grand Army of the Republic underscored. The professionals who had led these armies had strong roots in the continental French military experience from the tradition of the *levée en masse*, the teachings of Jomini, and the direct allied leadership of Frenchmen like Lafayette and later Foch—a powerful influence which fit into the Army's image of being the nucleus of an American continental strategic doctrine. In practice, however, America's strategic insularity had been guaranteed by naval power, primarily Britain's, until the U.S. Navy's expansion after 1900.

Aside from the colonies obtained in 1898, America's maritime empire was largely an informal one, keeping economic doors open not only in China but throughout the world by the presence of naval forces. The professional Army supported this strategic doctrine by defending and policing the borders and coasts of the country and its overseas possessions, while the Marines provided the Navy with its landing element during the interventions abroad. The Navy had no Grand Army or strong militia tradition to dramatize its role, although it tried to capitalize on the latter with the establishment of the short-lived Naval Militia from 1888, from whence emerged the Navy League of the United States in 1902 to educate the American people on the subtle applications of sea power. This strategic doctrine, however, imperfectly understood by the average American, was entirely realistic, for it met the needs of the country, and, except for the involvement in World War I, it continued throughout the 1920s and 1930s.[41]

Unilaterally, both Britain and the United States sought to maintain the same maritime *pax* after as before World War I and to control the seas in both world wars. But the same strategic circumstances of the international power relationships that had led to informal prewar cooperation forced an even closer union between them in both world wars and in the naval conferences of 1921–1922 and 1930 in which they recognized mutual naval parity—however begrudgingly.[42]

In World Wars I and II, the decisive commitment of the wartime U.S. Army in France—in 1918 and 1944—strengthened the image of an Army-led continental strategic doctrine; again, this was a naive notion in light of actual Anglo-American maritime policies and strategy. In both wars, the United States adopted Britain's traditional maritime strategy, or what British military pundit B. H. Liddell Hart called the "indirect approach." In this strategy, according to Sir Julian Corbett, Britain's foremost maritime strategist, the maritime nation (or coalition) must win command of the sea, as the Anglo-American navies did over the U-boat in 1917–1918 and 1943–1944, while a major allied continental army hammers away directly at the continental enemy; in 1914–1917 the Anglo-French-Russian armies, and in 1941–1944 the Red Army, fulfilled this function by wearing down the major German armies. At this point, when the enemy has lost the sea and has fallen back on the strategic defensive on land, "The intrusion of a small fresh force from the sea . . . may suffice to turn the scale . . .," with "the theater of operations assigned to it . . . such that in no probable event could it lose touch with the sea, nor could the enemy cut its lines of supply and retreat." Such was the strategic function of the American Expeditionary Force in 1918 and the Anglo-American Allied Expeditionary Forces in 1944. Both commitments achieved their objective, bringing on final German defeat, but only after having "indirectly" whittled down German arms from the sea (and air) and after the continental Allied armies had fought the major campaigns on land.[43]

Despite the initial American acceptance of the British indirect approach, and the Anglo-American agreement on the overall war aim to defeat Germany before Japan in World War II, events in 1941–1942 brought out important differences of strategic opinion among the high American commanders. Favoring the direct approach reminiscent of the "On to Richmond" continental policy of the Civil War, the Army, led by its Chief of Staff, Gen. George C. Marshall, tried to assume the leadership over the Anglo-American strategic policy by advocating an immediate cross-Channel attack, only to be stopped by the more realistic maritime doctrines of the British propounded by Prime Minister Winston Churchill.

In addition, the Navy had spent the 1920s and 1930s concentrating on Japan as the next enemy, with the result that it fully intended to wage maritime war in the Pacific, culminating in the blockade and defeat of Japan—with the Marines to take the smaller island bases and the Army the larger ones. Both the Army's commanders in Hawaii and Gen. Douglas MacArthur in the Southwest Pacific resented being subordinated to the Navy's strategic leadership; the former failed to break away, though MacArthur succeeded, only to then operate on a virtual shoestring until mid-1944. The Navy generally got its way in the Pacific, and when China did not materialize as the viable continental ally, Russia agreed in 1945 to provide its army for the purpose. Nevertheless, the U.S. Army insisted on a direct invasion of Japan proper in the continental strategic manner, but this became unnecessary when the maritime strategy succeeded.[45]

An additional factor in the popular American proclivity for the direct approach in the continental mode lay in the professed doctrines of strategic aerial bombing. Again the influence was continental, from Gen. Billy Mitchell's experiences in France in 1918 and especially the impact of the theories of Italian General Giulio Douhet at the Air Corps Tactical School from the mid-1920s. Douhet had argued that the strategic bomber could destroy industrial targets and civilian morale, thus substituting air forces as the decisive hitting element over armies—and navies, which Mitchell liked to stress. Douhet referred only to Europe, where during World War II British Bomber Command and the U.S. Army Air Forces attempted to live up to Douhet's theories, but in vain. According to Anthony Verrier, ". . . all the strategic novelty claimed for air power was rendered null and void" in its failure to destroy Germany's air forces, industry, or civilian morale. But as an adjunct of maritime strategy, the bombers (and defensive fighters) operating from insular England admirably succeeded, assisting to achieve the sea and air command that helped defend the English coast and sea lanes by constant aerial battles and by destroying U-boats under construction. The bombers also weakened the German economy, seriously interdicted German lines of communications especially prior to and after the D-Day assault, and forced Hitler onto the strategic defensive "by robbing his armies of air cover" in favor of home defense, a change which "most obviously benefitted the Russian armies" pressing into Germany from the East. The American strategic bombers thus served properly as a subordinate force within the U.S. Army, acting as the ultimate extension of long-range artillery.[46]

In the Pacific—of which Douhet had said nothing—Navy, Army, and Marine Corps air forces served as virtual extensions of the naval gun, isolating island battlefields from interference by the Japanese fleet, air forces, and seaborne troop reinforcements and providing air cover and close air support for the assault forces. The strategic bombers which pounded Japanese industry and cities in 1945 inflicted much damage but were most decisive as a component of maritime strategy in their massive aerial mining campaign of the general air-sea blockade which defeated Japan.

Despite these successes, wartime Air Force planners for the postwar world "developed a force structure that ignored every [major] air lesson of World War II" in order to achieve service autonomy based on the strategic bombardment mission of a basically anti-Soviet Russian continental strategy. Propagandizing that strategic air power had been the chief agent of victory in Europe and the Pacific, and reinforced by the awesomeness of the atomic bombs, the postwar bomber advocates got their way in a separate Air Force in 1947, leading to a fierce interservice battle in which the Navy tried to share such a strategic mission. Before long, however, the growth of nuclear and thermonuclear weapons reached the point in the 1950s—certainly by the time of the Cuban missile confrontation of 1962—that placed them into a special political category outside the normal context of military doctrines and strategy. Put another way, the use of such "absolute weapons" made them prohibitively dangerous ("unthinkable") as practical weapons of war. Their real function has remained one of political deterrence, but beyond that they have outgrown any actual military utility that they might have once had. Even their possible tactical use invites the very real probability of escalation into a general thermonuclear exchange. Whether land- or sea-based, such aircraft- or missile-launched bombs simply do not figure in the application of traditional continental or maritime strategy. Unfortunately, the illusion that they offered a strategic panacea led the Air Force and Navy to allow their tactical air forces that had proved so effective in World War II to degenerate, as painfully revealed in Korea and Vietnam.[47]

While the United States was professing a nuclear war-oriented strategy, it was in practice employing a maritime strategy which has been given the strategic label of *Pax Americana* to the era since 1945. Broadly speaking, this postwar *pax* has been merely a renewal of the same strategic umbrella that had protected Anglo-American territorial and commercial security during the nineteenth century, except that the relative roles played by Britain and the United States have been reversed. This became obvious in the North Atlantic and European waters from 1947, when Britain admitted its strategic prostration, an admission that surprised American strategic planners who had been counting on British leadership there. The nature of the American "empire" (largely informal and economic) did not change—the Philippines even obtained their political independence in 1946—while Britain's (and France's and Holland's, too) did (the former colonies breaking away peaceably or violently over the next twenty years). But economic spheres of influence remained, with the United States becoming the undisputed policeman of the Western Pacific following the collapse of the European and Japanese empires there and the rise of Communist China. American reluctance to assume such a global maritime commitment was compounded by the naive belief that a monolithic worldwide communist revolution was being directed by Soviet Russia, and that, therefore, nuclear weapons aimed at Russia should form the basis of American defense policy. Consequently, the American armed forces tooled for a possible thermonuclear World War III and found only doctrinal frustration in the limited wars of this maritime *pax*.[48]

However imperfectly understood, America's post-World War II maritime strategy has embodied the traditional functions of the Navy and Army, evolving through painful experience into the first real officially articulated oceanic policy, the Nixon Doctrine, in 1969. Under its nuclear umbrella and with a surface navy commanding the seas, the United States has sought major continental allies in Europe and Asia to provide the strong ground forces for any confrontation with Russia. Unfortunately, the NATO, CENTO, and SEATO alliances created between 1949 and 1956 to ring the periphery of the Eurasian continental land mass have succeeded less from strong Allied armies than from the American nuclear shield over Europe and direct American aid and involvement in the case of Asia. The Nixon Doctrine required that these allies contribute the ground forces while the United States controlled the sea and air. Success in the latter elements enabled the U.S. Navy to isolate the battlefields in peninsular Korea in 1950–1953 and insular Santo Domingo in 1965 and to blockade the island of Cuba from Russian seaborne missile cargoes in 1962.

But the United States has been unsystematic in asserting its position as a maritime power, refusing to utilize its superior Navy to blockade coastal China during the Korean War or coastal North Vietnam above 20°N. until the very last stages of the war there (1972–1973). The conventional forces of the U.S. Army, Air Force, and Marine Corps have performed their historic missions of providing the tactical effort ashore, in pacification and policing operations. Major weaknesses, however, have developed from the commitment of ground forces into the non-insular and strategically unfavorable country of South Vietnam, where "the west flank is open throughout its length to infiltration via Laos and Cambodia," and from an Army that from 1945 relied on conscripts to do the work of professional soldiers. Both of these problems were partially remedied (hopefully permanently) early in 1973 by the withdrawal of American ground forces from Vietnam and the ending of the draft.[49]

History and the Sea

In conclusion, the value of strategic history becomes apparent in the formulation of hypotheses for examining the continuum of America's (or any country's) military past, as well as for suggesting possible alternatives for future action. The episodic impressions of the "big wars" and of the apparent though inaccurate doctrinal superiority of large continental American armies and land-based air forces in the two world wars and the Cold War over the maritime elements in the minds of the American people and American historians can be eradicated only by the careful study of the complete overview of America's military past in practice as well as in theory. Until the recent scholarship noted above, and the whole new genre of literature now beginning to examine maritime strategic theory and the techniques of modern limited war,[50] such hypotheses have not been possible.

Closely related to this inability has been a recent preoccupation with the hardware of war, both in the popular mind and literature and in the thoughts and acts of strategy makers, which also has hindered progress in the role of historians to help comprehend more fully the place of armed conflict in human relations. Warns Raymond Aron, ". . . no amount of conceptualizations, schemes or scientific studies of quantifiable decisions (arms systems) are a substitute . . . for historical analysis of actual situations. . . . Strategic thought draws its inspiration each century, or rather at each moment of history, from the problems which events themselves pose."[51] Strategic history is part of this thought, offering possibilities for hypotheses heretofore not fully articulated in the evaluations of the continuum of American civilization.

REFERENCES

1. C. Vann Woodward, "The Age of Reinterpretation," address to the American Historical Association, printed in *The American Historical Review*, LXVI, 1 (Oct. 1960), 1–19, and reprinted as Service Center for Teachers of History Publication Number 35 (Washington, 1961). Woodward based his appeal on three factors: the end of America's "free security," the revolution in contemporary weapons technology, and the end of global European hegemony. Wesley Frank Craven made a similar appeal in 1959 to military professionals, namely, to "study military history, indeed all history, as no generation of military men have studied it before." W. Frank Craven, "Why Military History?" U.S. Air Force

American Strategic History and Doctrines

Academy Harmon Memorial Lecture in Military History Number One (1959), 9, 11. Correlli Barnett to the editor, *Times Literary Supplement* (20 Nov. 1969), 1338, in response to Michael Howard's fine essay, "The Demand for Military History," *TLS* (13 Nov. 1969), 1293–1294. See the sub-chapter, "Strategic History," in my *Command of the Sea: The History and Strategy of Maritime Empires* (New York, 1974), 9–12.

2. Walter Millis, *Military History*, Service Center Publication Number 39 (Washington, 1961), 17. That traditional forms of military history are no longer adequate is the theme of Russell F. Weigley's essay, "The End of Militarism," U.S. Air Force Academy Harmon Lecture Number Fifteen (1973).

3. Hans Delbrück, *Geschichte der Kriegskunst im Rahmen der politischen Geschichte* (Berlin, 1920), IV, no p., quoted by Peter Paret, "The History of War," in Felix Gilbert and Stephen R. Graubard, eds., *Historical Studies Today* (New York, 1972), 378. Analyses of the present aspects of war may, though not necessarily, also be useful in formulating an understanding of their past and of strategy in general, a methodology embodied in Delbrück's concept of *Sachkritik*. See Gordon A. Craig, "Delbrück: The Military Historian," in Edward Mead Earle, ed., *Makers of Modern Strategy* (1941: New York, 1966), 265–268. An excellent recent example of this methodology is John Shy, "The American Revolution: The Military Conflict Considered as a Revolutionary War," in Stephen G. Kurtz and James H. Huston, eds., *Essays on the American Revolution* (Chapel Hill, 1973), 121–156, in which he asks "whether the doctrines, the studies, and the general experience of 'revolutionary warfare' in the twentieth century provide some insight into the American Revolutionary War. The answer, with due caution and qualification, is yes." (123).

4. The best recent bibliographical essay of American military history, but carefully related to other countries and times, is Theodore Ropp, "War: From Colonies to Vietnam," in William H. Cartwright and Richard L. Watson, Jr., eds., *The Reinterpretation of American History and Culture* (Washington, 1973), 207–226.

5. Pioneering works in civil-military relations in America are Samuel P. Huntington, *The Soldier and the State* (New York, 1957); Paul Y. Hammond, *Organizing for Defense* (Princeton, 1961); and Harold Stein, ed., *American Civil-Military Decisions* (University, AL, 1963). In sociology, Morris Janowitz, *The Professional Soldier* (Glencoe, 1960); in psychology, John Shy, "The American Military Experience: History and Learning," *The Journal of Interdisciplinary History*, I, 2 (Winter 1971), 205–228, but see also L. Bramson and G. W. Goethals, eds., *War: Studies from Psychology, Sociology, and Anthropology* (New York, 1964); in economics, among many others, Charles J. Hitch and Roland N. McKean, *The Economics of Defense in the Nuclear Age* (Cambridge, MA, 1960); and, the most comprehensive interdisciplinary work of all, Quincy Wright, *A Study of War*, 2nd ed. (Chicago, 1965). On the philosophy and literature of war, see R. T. Ginsberg, ed., *The Critique of War* (Chicago, 1969), and Ernest Hemingway ed., *Men at War* (New York, 1942).

6. The leading classical scholars remain the German Karl von Clausewitz and the Briton Sir Julian Corbett. The Penguin edition of the former's 1832 work is a good condensation: Anatol Rapoport, ed., *On War* (Baltimore, 1968). Newly reprinted is Corbett's 1911 work, *Some Principles of Maritime Strategy* (Annapolis, 1972).

For a summary of these thinkers and their successors, see Michael Howard, "The Classical Strategists," in The Institute for Strategic Studies (Studies in International Security No. 14), *Problems of Modern Strategy* (London, 1970), 47–76. Howard himself may be regarded as a leading strategic historian today. See especially his *Studies in War and Peace* (New York, 1972), a collection of essays which includes a reprint of the above essay. The work of such universal historians as H. G. Wells, Arnold Toynbee, and W. H. McNeill are always useful for the shaping of new hypotheses, but other pertinent works, heavily influenced by the German scholarship of the Delbrück-Ranke era, are Ellen Churchill Semple, *Influences of Geographic Environment* (New York, 1911) and, published posthumously, Herbert Rosinski (Richard P. Stebbins, ed.), *Power and Human Destiny* (New York, 1965).

7. Theodore Ropp, "The Historical Development of Contemporary Strategy," U.S. Air Force Academy Harmon Lecture Number Twelve (1970), 5, 18, citing Thomas S. Kuhn, *The Structure of Scientific Revolutions* (Chicago, 1964). For sound examples of generational theory, again derived from the Delbrück-Ranke German historical school, see Fritz Redlich's treatments of American business entrepreneurs and German writers, *History of American Business Leaders* (Ann Arbor, 1940), 22ff. and "German Literary Expressionism and Its Publishers," *Harvard Library Bulletin*, XVIII, 2 (Apr. 1969), 143–168.

8. See Huntington, *Soldier and the State*, 143–221.

9. Paret, "The History of War," 380.

10. Shy, "The American Military Experience," 207. Though now a civilian, Shy is a graduate of the U.S. Military Academy, class of 1952.

11. Peter Paret and John Shy, *Guerrillas in the 1960's*, rev. ed. (New York, 1962), 3.

12. Russell F. Weigley, *The American Way of War* (New York, 1973), 477. Weigley had not progressed far beyond Millis, upon whose *Arms and Men* (New York, 1956, 1967) he heavily relied. Millis had also written twelve years before, "If there is one thing clear about the institution of war today, it seems to be that to continue to project its past history into the future is not merely to invite but to make certain a total catastrophe of civilization." *Military History*, 16. Craven challenged Millis' pessimism, also evident in *Arms and Men*, 7, in his "Why Military History?," 8–9.

13. The Royal Navy did not become an object of American resentment until the Revolution, specifically the bombardment of Falmouth, Maine, in 1775 and the subsequent operations along the coasts of New England and New York, after which American leaders, notably John Adams, tried to manufacture pre-Revolution anti-naval issues such as impressment. See Phillip S. Haffenden, "Community and Conflict: New England and the Royal Navy, 1689–1775" (unpublished paper, 12th Conference of the International Commission for Maritime History, Greenwich, England, July 1974); Neil R. Stout, *The Royal Navy in America, 1760–1775: A Study of Enforcement of British Colonial Policy in the Era of the American Revolution* (Annapolis, 1973); John J. Kelly, Jr., "The Struggle for American Seaborne Independence as Viewed by John Adams," (Ph.D. dissertation, University of Maine, 1973); Donald R. Yerxa, "Admiral Samuel Graves and the Falmouth Affair: A Case Study in British Imperial Pacification, 1775, (unpublished M.A. thesis, University of Maine, 1974); and Ira D. Gruber, *The Howe Brothers and the American Revolution* (New York, 1972). For a balanced

view of the British Army's contribution to the colonists' antagonism toward imperial rule, see John Shy, *Toward Lexington: The Role of The British Army in the Coming of the American Revolution* (Princeton, 1965).

14. Millis, *Arms and Men*, 6. A recent example of general military history ignoring the Navy is Marcus Cunliffe, *Soldiers and Civilians: The Martial Spirit in America, 1775-1865* (Boston, 1968), 27n.

15. Ibid.; Harold and Margaret Sprout, *The Rise of American Naval Power, 1776-1918*, rev. ed. (Princeton, 1939, 1967), v–xiii; Alfred Thayer Mahan, *The Influence of Sea Power upon History, 1660-1783* (Boston, 1890). Bernard Brodie, *Sea Power in the Machine Age* (Princeton, 1941), complemented the Sprouts' work with the technological dimension.

16. Harold and Margaret Sprout, *Toward a New Order of Sea Power: American Naval Policy and the World Scene, 1918-1922*, 2nd ed. (Princeton, 1943; New York, 1969), 289.

17. For the first hypothesis, see my *Command of the Sea*, 3 ff. The genesis of my categorization lay in Corbett's *Some Principles of Maritime Strategy*, in which (38–48, passim) Corbett criticized Clausewitz for being exclusively continental oriented and thus ignorant of maritime strategy, a gap which Corbett then proceeded to fill. The coming of total war in 1914 just three years after Corbett's book had appeared tended to bury his hypotheses of limited war and other less episodic uses of sea power. Also, some bad guesses about navies in a modern naval war further undermined the credibility of his other and more enduring theories.

18. Woodward, "The Age of Reinterpretation," 2.

19. This thesis is developed more fully in my essay, "The British Strategic Inheritance in American Naval Policy, 1775-1975," in Benjamin W. Labaree, ed., *The Atlantic World of Robert G. Albion* (Middletown, CN, 1975), 216-247.

20. See ibid.; my *Command of the Sea*; Corbett, *Some Principles of Maritime Strategy*; Vice Admiral Sir Peter Gretton, *Maritime Strategy* (New York 1965); and James Cable, *Gunboat Diplomacy: Political Applications of Limited Naval Force* (London, 1971). For Britain, see especially Gerald S. Graham, *The Politics of Naval Supremacy* (Cambridge, Eng., 1965); Captain S. W. Roskill, *The Strategy of Sea Power* (London, 1962); Donald M. Schurman, *The Education of a Navy* (Chicago, 1965); Correlli Barnett, *Britain and Her Army, 1509-1970* (London, 1970); and Jay Luvaas, *The Education of an Army, 1815-1940* (Chicago, 1964).

21. See John U. Nef, *War and Human Progress* (1950; New York, 1963), 328, 376-377.

22. Robert Greenhalgh Albion, with Jennie Barnes Pope, *Sea Lanes in Wartime: The American Experience, 1775-1945*, 2nd enl. ed. ([Hamden, CN], (1968), 34; Albion, William A. Baker and Benjamin W. Labaree, *New England and the Sea* (Middleton, CN, 1973), 64.

23. Douglas Edward Leach, *Arms for Empire: A Military History of the British Colonies in North America, 1607-1763* (New York, 1973); Darrett Bruce Rutman, "A Militant New World, 1607-1640: America's First Generation, Its Martial Spirit, Its Tradition of Arms, Its Militia Organization, Its Arms" (Ph.D. dissertation, University of Virginia, 1959); William Alfred Foote, "The American Independent Companies of the British Army, 1664-1764," (Ph.D. dissertation, U.C.L.A., 1966) and David Richard Millar, "The Militia, the Army, and Inde-

History and the Sea

pendency in Colonial Massachusetts," (Ph.D. dissertation, Cornell University, 1967). John Shy's call for "A New Look at Colonial Militia," *William and Mary Quarterly*, 3rd ser., XX, 2 (Apr. 1963), 175–185, suggests regional differences, but in strategic history militia is militia, as opposed to a large professional standing army.

24. Nathan Miller, *Sea of Glory* (New York, 1974); Sprouts, *American Naval Power*, 7ff.

25. Frederic H. Hayes, "John Adams and American Sea Power," *American Neptune*, XXV, 1 (Jan. 1965), 35-45; Kelly, "American Seaborne Independence," 109–110, 153, 304, 308; Howard I. Chapelle, *The History of the American Sailing Navy* (New York, 1949), 38–39, 120–121, 126, 177–178, 194, 234–235; and the Sprouts, *American Naval Power*, 16ff., 25ff., 50ff. Joseph George Henrich, "The Triumph of Ideology: The Jeffersonians and the Navy, 1779–1807" (Ph.D. dissertation, Duke University, 1971) shows a virtual non-naval policy by Jefferson until he opted for the gunboat over the frigate after the *Chesapeake* affair of 1807.

26. Don Higginbotham, *The War of American Independence: Military Attitudes, Policies, and Practice, 1763–1789* (New York, 1971), 4, 94, 432, which supersedes all other works on the subject, though still useful is Millis, *Arms and Men*, 13 ff. A recent treatment of regional guerrilla operations is Russell F. Weigley, *The Partisan War: The South Carolina Campaign of 1780–1782* (Columbia, SC, 1970).

27. Richard Henry Kohn, "The Federalists and the Army: Politics and the Birth of the Military Establishment, 1783–1795" (Ph.D. dissertation, University of Wisconsin, 1968); John K. Mahon, *The American Militia: Decade of Decision, 1789–1800* (Gainesville, 1960). For an overview of the period 1775–1815, see Weigley, *American Way of War*, 3ff.

28. The definitive account is John K. Mahon, *The War of 1812* (Gainesville, 1972), but see the Sprouts, *American Naval Power*, 73ff.

29. Harry L. Coles, *The War of 1812* (Chicago, 1965), 271. See also Millis, *Arms and Men*, 71, 99, passim.

30. C. J. Bartlett, *Great Britain and Sea Power, 1815–1853* (Oxford, 1963); Emanuel Raymond Lewis, *Seacoast Fortifications of the United States: An Introductory History* (Washington, 1970); Sprouts, *American Naval Power*, 110ff., 127ff.; James A. Field, Jr., *America and the Mediterranean World, 1776–1882* (Princeton, 1969); Grace Fox, *British Admirals and Chinese Pirates, 1832–1869* (London, 1940); Robert Erwin Johnson, *Thence Round Cape Horn: The Story of United States Naval Forces on Pacific Station, 1818–1923* (Annapolis, 1963); Curtis Talmon Henson, Jr., "The United States Navy and China, 1839–1861" (Ph.D. dissertation, Tulane University, 1961); Benjamin Franklin Gilbert, "Naval Operations in the Pacific, 1861–1866" (Ph.D. dissertation, University of California, Berkeley, 1951); Kenneth Bourne, *Britain and the Balance of Power in North America, 1815–1908* (Berkeley, 1967); Barry M. Gough, *The Royal Navy and the Northwest Coast of North America, 1810–1914* (Vancouver, 1971); William A. Morgan, "Sea Power in the Gulf of Mexico and the Caribbean during the Mexican and Colombian Wars of Independence, 1815–1830" (Ph.D. dissertation, University of Southern California, 1969); Maury Davison Baker, Jr., "The United States and Piracy during the Spanish-American Wars of Indepen-

dence" (Ph.D. dissertation, Duke University, 1946); Richard Carl Froehlich, "The United States Navy and Diplomatic Relations with Brazil, 1822–1871" (Ph.D. dissertation, Kent State University, 1971); Edward Warner Billingsley, *In Defense of Neutral Rights: The United States Navy and the Wars of Independence in Chile and Peru* (Chapel Hill, 1967); Richard Wellington Turk, "Strategy and Foreign Policy: The United States Navy in the Caribbean, 1865–1913" (Ph.D. dissertation, Fletcher School of Law and Diplomacy, 1968); Lance C. Buhl, "The Smooth Water Navy: American Naval Policy and Politics, 1865–1876" (Ph.D. dissertation, Harvard University, 1968); and Kenneth J. Hagan, *American Gunboat Diplomacy and the Old Navy, 1877–1889* (Westport, CN, 1973). On anti-slaving operations, see Christopher Lloyd, *The [Royal] Navy and the Slave Trade* (New York, 1949), and Earl E. McNeilly, "The United States and the Suppression of the West African Slave Trade, 1819–1862" (Ph.D. dissertation, Case Western Reserve, 1973).

31. Stephen E. Ambrose, *Duty, Honor, Country: A History of West Point* (Baltimore, 1966), 66ff.; Thomas E. Griess, "Dennis Hart Mahan: West Point Professor and Advocate of Military Professionalism, 1830–1871" (Ph.D. dissertation, Duke University, 1968).

32. Francis Paul Prucha, *The Sword of the Republic: The United States Army on the Frontier, 1783–1846* (New York, 1969), which begins with the statement, "The regular army of the United States owed its existence to the American frontier." Also Prucha, *Broadax and Bayonet: The Role of the United States Army in the Development of the Northwest, 1815–1860* ([Madison], 1953); Robert M. Utley, *Frontiersmen in Blue: The United States Army and the Indian, 1848–1865* (New York, 1967); John K. Mahon, *History of the Second Seminole War, 1835–1842* (Gainesville, 1967); George E. Buker, *Swamp Sailors: Riverine Warfare in the Everglades, 1835–1842* (Gainesville, 1975); (Ph.D. dissertation, University of Florida, 1969); Robert G. Athearn, *Forts of the Upper Missouri* (Englewood Cliffs, NJ, 1967); George Rogers Taylor, *The Transportation Revolution, 1815–1860* (New York, 1951); and K. Jack Bauer's two works, with the many episodes of anti-guerrilla operations, *The Mexican War* (New York, 1974), and *Surfboats and Horse Marines: U.S. Naval Operations in the Mexican War, 1846–1848* (Annapolis, 1969). The later Indian wars are covered in Robert M. Utley, *Frontier Regulars: The United States Army and the Indians, 1866–1891* (New York, 1974). See also Weigley, *American Way of War*, 59ff.

33. "The Demand for Military History," 1294.

34. John D. Hayes, "Sea Power in the Civil War," *U.S. Naval Institute Proceedings*, 87, 11 (Nov. 1961), 60–69; Theodore Ropp, "Anacondas Anyone?" *Military Affairs*, XXVII, 2 (Summer 1963), 71–76; John B. Heffernan, "The Blockade of the Southern Confederacy: 1861–1865," *The Smithsonian Journal of History*, II, 4 (Winter 1967–1968), 23–44; Samuel R. Bright, Jr., "Confederate Coast Defense," (Ph.D. dissertation, Duke University, 1961); William N. Still, Jr. *Confederate Shipbuilding* (Athens, GA, 1969), and the Sprouts, *American Naval Power*, 151ff. See n. 44 below.

35. Shy, "The American Military Experience," 219.

36. Utley, *Frontiersmen in Blue*. For Mosby, whom he calls a "semi-regular," and other guerrillas, see Ethelbert Courtland Barksdale, "Semi-Regular and Irregular

Warfare in the Civil War," (Ph.D. dissertation, University of Texas, 1941), 66ff, 88ff. Also, Clark G. Reynolds, "The Civil and Indian War Diaries of Eugene Marshall, Minnesota Volunteer, 1861–1866" (unpublished M.A. thesis, Duke University, 1963), which treats the Sioux and Tennessee operations in some detail; Richard S. Brownlee, *Gray Ghosts of the Confederacy: Guerrilla Warfare in the West, 1861–1865* (Baton Rouge, 1958); Carl W. Breihan, *Quantrill and Civil War Guerrillas* (Denver, 1959); and Gary Norton, "The Mississippi Marine Brigade, 1862–1865" (uncompleted M.A. thesis, University of Maine).

37. Sprouts, *American Naval Power*, 165ff.; James E. Sefton, *The United States Army and Reconstruction, 1865–1877* (Baton Rouge, 1967); Jerry Marvin Cooper, "The Army and Civil Disorder: Federal Military Intervention in American Labor Disputes, 1877–1900" (Ph.D. dissertation, University of Wisconsin, 1971); Joseph John Holmes, "The National Guard of Pennsylvania: Policemen of Industry, 1865–1905" (Ph.D. dissertation, University of Connecticut, 1971); Richard Allen Andrews, "Years of Frustration: William T. Sherman, the Army, and Reform, 1869–1883" (Ph.D. dissertation, Northwestern University, 1968); Richard H. Zeitlin, "Brass Buttons and Iron Rails: The United States Army and American Involvement in Mexico, 1868–1881" (Ph.D. dissertation, University of Wisconsin, 1973); and Stephen E. Ambrose, *Upton and the Army* (Baton Rouge, 1964). For a treatment of the occasional use of the Army as policeman over civil disorders at home from the 1790s to the 1960s, see Robin Higham, ed. *Bayonets in the Streets: The Use of Troops in Civil Disturbances* (Lawrence, KS, 1969; rev. Hamden, CN, 1976).

38. Quoted in Richard D. Challener, *Admirals, Generals, and American Foreign Policy, 1898–1914* (Princeton, 1973), 26. See passim, also John A. S. Grenville and George Berkeley Young, *Politics, Strategy, and American Foreign Policy, 1873–1917* New Haven, 1966), 270ff., 307; Nef, *War and Human Progress*, 368; The Sprouts, *American Naval Power*, 188, 213, 241–247; and Weigley, *American Way of War*, 167ff. The new American imperialism was anticipated within the U.S. Navy in the 1880s. See Hagan, *American Gunboat Diplomacy;* Ronald Harvey Spector, " 'Professors of War'; The Naval War College and the Modern American Navy," (Ph.D. dissertation, Yale University, 1967); and Robert Seager II, "Ten Years before Mahan: The Unofficial Case for the New Navy, 1880–1890," *Mississippi Valley Historical Review*, 40 (1953–1954), 491–512. On the American naval and diplomatic support of British policy in South Africa by the 1880s and 1890s, see Thomas J. Noer, "Commodore Robert W. Shufeldt and America's South African Strategy," *American Neptune*, XXXIV, 2 (Apr. 1974), 86–88. The British, reacting to American imperialism, generally "welcomed the extension of Anglo-Saxon law and order to hitherto ill-governed or undeveloped countries." Basil Collier, *The Lion and the Eagle: British and Anglo-American Strategy, 1900–1950 (New York, 1972)*, 19; Arthur J. Marder, *From the Dreadnought to Scapa Flow: The Royal Navy in the Fisher Era, 1904–1919*, 5 vols., I: *The Road to War, 1904–1914* (London, 1961), 40–42, 124, 125, 183–184; and Colonel Roger Willock, USMCR, "Gunboat Diplomacy: Operations of the [British] North America and West Indies Squadron, 1875–1915," *American Neptune*, XXVIII, 1 (Jan. 1968), 5–30, and XXVIII, 2 (Apr. 1968), 85–112.

39. Challener, *Admirals and Generals*, 16ff., 34ff., 81ff., 111ff., 288–315, 323ff., 379ff.; the two books by William Reynolds Braisted, *The United States Navy in*

the Pacific, 1897–1909 (Austin, 1958), and *1909–1922* (Austin, 1971); the three articles by Seward W. Livermore, "The American Navy as a Factor in World Politics, 1903–1913," *American Historical Review*, LXIII, 4 (July 1958), 863–879, "American Strategy Diplomacy in the South[eastern] Pacific, 1890–1914," *Pacific Historical Review*, XII, 1 (Mar. 1943), 33–51, and "Battleship Diplomacy in South America: 1905–1925," *Journal of Modern History*, XVI, 1 (Mar. 1944), 31–48; the Sprouts, *American Naval Power*, 251–285; Grenville and Young, *Politics and Strategy*, 301–306; Daniel Joseph Costello, "Planning for War: A History of the General Board of the Navy, 1900–1914" (Ph.D. dissertation, Fletcher School of Law and Diplomacy, 1968); and Colonel Robert Debs Heinl, Jr., USMC, *Soldiers of the Sea: The United States Marine Corps, 1775–1962* (Annapolis, 1962), 111ff.

40. Ibid., 23–26. Graham A. Cosmas, *An Army for Empire: The United States Army in the Spanish-American War* (Columbia, MO, 1971), 80–82, 102–107, 121ff., shows that even during the war with Spain the Army preferred victory in Cuba through naval blockade alone and yielded to landing operations only at the insistence of President McKinley and the Navy, which disdained the rigors of prolonged blockade in the tropics and the inability to maneuver while on station with the Spanish fleet still at large. Also, Donald Smythe, *Guerrilla Warrior: The Early Life of John J. Pershing* (New York, 1973); James William Pohl, "The General Staff and American Military Policy: The Formative Period, 1898–1917" (Ph.D. dissertation, University of Texas, 1967); Jack Constant Lane, "Leonard Wood and the Shaping of American Defense Policy, 1890–1920," (Ph.D. dissertation, University of Georgia, 1963); David F. Healy, *The United States in Cuba, 1898–1902: Generals, Politicians, and the Search for Policy* (Madison, 1963); Allan Reed Millett, *The Politics of Intervention: The Military Occupation of Cuba, 1906–1909* ([Columbus,] 1968); John Morgan Gates, *Schoolbooks and Krags: The United States Army in the Philippines, 1898–1902* (Westport, CN, 1973); Peter Gordon Gowing, "Mandate in Moroland: The American Government of Muslim Filipinos, 1899–1920" (Ph.D. dissertation, Syracuse University, 1968); George W. Jornacion, "The Time of the Eagles: United States Army Officers and the Pacification of the Philippine Moros, 1899–1913," (Ph.D. dissertation, University of Maine, 1973); George Yarrington Coats, "The Philippine Constabulary: 1901–1907," (Ph.D. dissertation, Ohio State University, 1968); and Clarence C. Clendenen, *Blood on the Border: The United States Army and The Mexican Irregulars* (New York, 1969), 152ff. for the period 1914–1919.

41. On American geographic insularity in the early 20th century, see Nef, *War and Human Progress*, 376–377. Kevin R. Hart, "Towards a Citizen Sailor: The History of The Naval Militia Movement, 1888–1898," *American Neptune*, XXXIII, 4 (Oct. 1973), 258–279, shows that the Naval Militia reached its zenith during the Spanish-American War and steadily declined thereafter until being generally replaced by the U.S. Naval Reserve after 1915, but officially in 1918. See also Armin Rappaport, *The Navy League of the United States* (Detroit, 1962). On U.S. Marine Corps interventions in the Caribbean during the 1920s, see Heinl, *Soldiers of the Sea*, 231ff., and Neill Macauly, *The Sandino Affair* (Chicago, 1967). For the American Army occupation of post-World War I Germany see Keith LeBaun Nelson, "The First American Military Occupation in

Germany, 1918–1923," (Ph.D. dissertation, University of California, Berkeley, 1965). The presence of the U.S. Army in the Philippines culminated in the appointment of Gen. Douglas MacArthur as "field marshal" of the Philippine Army in 1936. D. Clayton James, *The Years of MacArthur, I*: 1880–1941 (Boston, 1970), 479ff. For the Navy in China, see Kemp Tolley, *Yangtze Patrol: The U.S. Navy in China* (Annapolis, 1971), covering the years 1898–1941. For the interwar Army in general, see Robert A. Miller, "The United States Army during the 1930's" (Ph.D. dissertation, Princeton University, 1973).

42. For the gradual rise of the United States over Great Britain in Anglo-American strategy making, see Collier, *Lion and the Eagle*. For the interwar navies, see Stephen Roskill, *Naval Policy Between the Wars:, Vol. I The Period of Anglo-American Antagonism, 1919–1929* (London, 1968), and the forthcoming Vol. II for the 1930s. The failure of the United States to build up to treaty strength has been laid to Congress and not the treaties in several recent dissertations: Ernest Andrade, Jr., "United States Naval Policy in the Disarmament Era, 1921–1937" (Michigan State University, 1966); Meredith William Berg, "The United States and the Breakdown of Naval Limitation, 1934–1939," (Tulane University, 1966); John C. Walter, "The Navy Department and the Campaign for Expanded Appropriations, 1933–1938" (University of Maine, 1972); Robert Levine, "The Politics of Naval Rearmament, 1930–1938" (Ph.D. dissertation, Howard University, 1972); and Calvin William Enders, "The Vinson Navy," (Michigan State University, 1970). See also John R. M. Wilson, "The Quaker and the Sword: Herbert Hoover's Relations with the Military," *Military Affairs*, XXXVIII, 2 (Apr. 1974), 42–44 passim.

43. Corbett, *Some Principles of Maritime Strategy*, 61–73, his examples being Britain in the Seven Years and Napoleonic wars. Also, B. H. Liddell Hart, *Strategy*, 2nd rev. ed. (New York, 1967), 18–19, passim, including references to his first articulation of "the strategy of indirect approach" in 1929. For the U.S. Army and Navy in World War I, see Edward M. Coffman, *The War to End All Wars* (New York, 1968); Harvey A. DeWeerd, *President Wilson Fights His War* (New York, 1968); and Weigley, *American Way of War*, 192ff., For unilateral American Naval policy before and during World War I, see Warner R. Schilling, "Admirals and Foreign Policy, 1913–1919" (Ph.D. dissertation, Yale University, 1953), and the Sprouts, *American Naval Power*, 344–347, 359ff. For Anglo-American naval relations before and during World War I, the best sources are Braisted, *USN in the Pacific, 1909–1922*, 171 ff., 343ff. and David F. Trask, *Captains and Cabinets: Anglo-American Naval Relations, 1917–1918* (Columbus, MO, 1972). For an overview of the American military effort in Europe in World War II, see Charles B. MacDonald, *The Mighty Endeavor: American Armed Forces in the European Theater in World War II* (New York, 1969).

44. Louis Morton, "Germany First: The Basic Concept of Allied Strategy in World War II," in Kent Roberts Greenfield, ed., *Command Decisions* (Washington, 1960), 11–47; Greenfield, *American Strategy in World War II* (Baltimore, 1963), 24 ff.; Weigley, *American Way of War*, 312ff.; Howard, *Studies in War and Peace*, 192; and Richard W. Steele, *The First Offensive, 1942: Roosevelt, Marshall and the Making of American Strategy* (Bloomington, 1973). Marshall's biographer observes, "In the study of strategy and tactics, Marshall suffered the same handicap as most American commanders in the period between the great

wars: . . . the lack of any need to think in terms of international relations." Forrest
C. Pogue, *George C. Marshall*, I: *Education of a General, 1880–1939* (New York,
1963), 347. See also Vol. II: *Ordeal and Hope, 1939–1942* (New York, 1966),
158–159, 266–267, 273, 321ff., 376–381; and Vol. III: *Organizer of Victory,
1943–1945* (New York, 1973), 173–176, 195–200, 242–244, 253–254, 334, 443,
526–528. For Eisenhower, see Stephen E. Ambrose, *The Supreme Commander:
The War Years of General Dwight D. Eisenhower* (Garden City, 1970), 30–36,
64–69, 359–362, passim. All the above works and those below draw heavily upon
the extensive Anglo-American official histories and occasional autobiographies.
One might hypothesize that, strategically, in the Civil War the Union had—in
effect—ultimately defeated the Confederacy using an indirect maritime approach; the Army of the Potomac provided the main continental striking arm in
Virginia against the principal Confederate ground forces while the blockade and
the Western armies wore down rebel resistance on the periphery.

45. Weigley, *American Way of War*, 242ff., 269ff. and my *The Fast Carriers: The
Forging of an Air Navy* (New York, 1968). On the development of Navy, Marine
Corps, and Army amphibious doctrines in the war against Japan, see Vice
Admiral George Carroll Dyer, USN (Ret.), *The Amphibians Came to Conquer:
The Story of Admiral Richmond Kelly Turner*, 2 vols. (Washington, 1972); Jeter
A. Isely and Philip A. Crowl, *The U.S. Marines and Amphibious War* (Princeton,
1951); and Vice Admiral Daniel E. Barbey, USN (Ret.), *MacArthur's Amphibi-
ous Navy: Seventh Amphibious Force Operations, 1943–1945* (Annapolis, 1969).

46. Anthony Verrier, *The Bomber Offensive* (New York, 1968), 18. For the develop-
ment of American air doctrines during and between both world wars, see I. B.
Holley, Jr., *Ideas and Weapons: Exploitation of the Aerial Weapon by the
United States During World War I* (New Haven, 1953); Raymond Richard
Flugel, "United States Air Power Doctrine: A Study of the Influence of William
Mitchell and Giulio Douhet at the Air Corps Tactical School, 1921–1935," (Ph.D.
dissertation, University of Oklahoma, 1965); Edward Warner, "Douhet,
Mitchell, Seversky: Theories of Air Warfare," in Earle, *Makers of Modern
Strategy*, 485–503; Major Alfred F. Hurley, USAF, *Billy Mitchell: Crusader for
Air Power* (New York, 1964); Reynolds, *The Fast Carriers*; Robin Higham, *Air
Power: A Concise History* (New York, 1973); Greenfield, *American Strategy*,
85ff.; Millis, *Arms and Men*, 283; and Weigley, *American Way of War*, 223ff.

47. Perry McCoy Smith, *The Air Force Plans for Peace, 1943–1945* (Baltimore,
1970), 16–17, 25–26, 28, 35, passim. For the period 1945–1960, see Samuel P.
Huntington, *The Common Defense* (New York, 1961), and for these and
subsequent years, Paul Y. Hammond, "Super Carriers and B–36 Bombers:
Appropriations, Strategy and Politics," in Stein, ed., *American Civil-Military
Decisions*, 465–564; Reynolds, *Command of the Sea*, 545ff.; Bernard Brodie,
Strategy in the Missile Age (Princeton, 1959, 1965); George E. Lowe, *The Age of
Deterrence* (Boston, 1964); Chalmers M. Roberts, *The Nuclear Years: The Arms
Race and Arms Control, 1945–1970* (New York, 1970); R. Earl McClendon,
Army Aviation, 1947–1953 (Air University Documentary Study, 1954); and
Weigley, *American Way of War*, 382ff., 399ff. "The problem of deterrence has
grown out of traditional and historic military problems. . . . The time may not be
far off, however, when it will outgrow them altogether: when its difficulties will
be purely technical and political, with very little military content; very little, that

is, of tactics and strategy left at all." Howard, *Studies in War and Peace*, 151. This was written in 1962, the same year that Nels A. Parson, Jr., *Missiles and the Revolution in Warfare* (Cambridge, MA, 1962) argued the same case.

48. On Anglo-American post-World War II strategic relations, see Vincent Davis, *Postwar Defensive Policy and the U.S. Navy, 1943–1946* (Chapel Hill, 1966), 29–30, 91–92 passim; R. N. Rosecrance, *Defense of the Realm: British Strategy in the Nuclear Epoch* (New York, 1968); B. B. Schofield, *British Sea Power: Naval Policy in the Twentieth Century* (London, 1967), 217 ff.; William James Crowe, Jr., "The Policy Roots of the Modern Royal Navy, 1946–1963" (Ph.D. dissertation, Princeton University, 1965). See also Lee S. Houchins, "American Naval Involvement in the Chinese Civil War, 1945–1949" (Ph.D. dissertation, American University, 1971).

49. J. Lawton Collins, *War in Peacetime: The History and Lessons of Korea* (Boston, 1969), 388, in which General Collins contrasts the Korean and Vietnam wars; Weigley, *American Way of War*, 382ff., 441 ff.; Rear Admiral John D. Hayes, USN (Ret.), "Patterns of American Sea Power, 1945–1956: Their Portents for the Seventies," *U.S. Naval Institute Proceedings*, 96, 5 (May 1970), 337–352, and "Sea Power and Sea Law," *Proceedings*, 90, 5 (May 1964), 60–67; James A. Field, Jr., *History of United States Naval Operations: Korea* (Washington, 1962); Commander Andrew J. Valentine, USN, "R$_x$: Quarantine," *Proceedings*, 89, 5 (May 1963), 38–50, on the Cuban blockade; Carl Q. Christol and Charles R. Davis (and Quincy Wright), "Maritime Quarantine: The Naval Interdiction of Offensive Weapons . . . to Cuba, 1962," *American Journal of International Law*, 57 (July 1963), 525–565; Commander R. L. Schreadly, USN, "The Naval War in Vietnam, 1950–1970," *Proceedings*, 97, 5 (May 1971), 180–209; Commander James A. Barber, USN, "The Nixon Doctrine and the Navy," *Naval War College Review*, XXIII, 10 (June 1971), 5–15; and Earl C. Ravenal, "The Nixon Doctrine and Our Asian Commitments," *Foreign Affairs*, 49, 2 (Jan. 1971), 201–217. On U.S. military aid to its allies, see Harold A. Hovey, *United States Military Assistance: A Study of Policies and Practices* (New York, 1965). A concise case study of American military advisers is Major Robert K. Sawyer (Walter G. Hermes, ed.), *Military Advisors in Korea: KMAG in Peace and War* (Washington, 1962).

50. See n.20 above; also Shy, "The American Military Experience,"; L. W. Martin, *The Sea in Modern Strategy* (New York, 1967); Michael F. H. Dennis, "The Role of Navies in Limited War" (Ph.D. dissertation, University of Minnesota, 1971); and Morton H. Halperin, *Limited War in the Nuclear Age* (New York, 1963). Two excellent recent studies examining the possible doctrines of the U.S. Army in the post-Vietnam *Pax Americana* are Colonel Wesley W. Yale, USA (Ret.), General I. D. White, USA (Ret.), and General Hasso E. von Manteuffel, German Army (Ret.), *Alternative to Armageddon: The Peace Potential of Lightning War* (New Brunswick, NJ, 1970), and John R. Galvin, *Air Assault: the Development of Airmobile Warfare* (New York, 1969).

51. Raymond Aron (tr. by J. E. Gabriel), "The Evolution of Modern Strategic Thought," in The Institute for Strategic Studies, *Problems of Modern Strategy*, 25, 45.

6

THE CONTINENTAL STATE UPON
THE SEA: IMPERIAL JAPAN

The continental state, as defined in chapters 1 and 2, has been the opposite of the maritime thalassocracy in every essential of its national life, especially in its activities upon the sea. Authoritarian governments like sixteenth-century Spain and Portugal, Bourbon France, and even Wilhelmian Germany developed active waterborne trade; seized colonies from weak, non-Western native peoples; and even built splendid warships. But this combination of ingredients for sea power, as defined by Mahan, never overcame the straitjackets of one-man autocratic political rule, a centralized economy, or army supremacy to counter the thalassocracies in imperial rivalries or naval wars.

Twentieth-century Imperial Japan—like its contemporary Nazi Germany—fell into this same category of continental state, although this is a difficult reality for Americans to accept, dazzled as they were and still are by the Pearl Harbor attack and Japanese naval prowess in World War II. Viewing Japan as a continental power, however, makes its defeat in the Pacific war more readily understandable. This chapter, originally titled "The Continental Strategy of Imperial Japan," appeared in the *U.S. Naval Institute Proceedings* in August 1983.

The Soviet Union is a continental power in every sense of the term—geographic, political, economic, strategic—and yet for the

Reprinted from *Proceedings* with permission: Copyright © 1983 U.S. Naval Institute.

past two decades it has produced a navy of awesome proportions. In looking for possible historical parallels against which to measure Soviet intentions and capabilities at sea, analysts have only to study the example of the Japanese Empire. Unfortunately, because of Japan's island position (and resulting false geopolitical comparisons with the British Isles), and because of the great sea battles of World War II, historians have tended to regard Imperial Japan as having been a maritime state. Nothing could be further from the truth.[1]

A maritime state may be defined as a nation whose interests are centered on overseas trade, possessions, and dependencies and not on any continental landmass. A strong middle class of capitalists directs a primarily manufacturing and industrial economy and a large, privately owned merchant marine. This free enterprise ethic is shared with free social institutions to shape a liberal form of government, that is, a democracy. The maritime nation depends upon its navy as the senior armed service to control not only the surrounding home waters but the vast reaches of its oceanic trade routes as well. The classic historical models of maritime states are ancient Athens, Renaissance Venice and the Netherlands, and modern Great Britain and the United States.[2]

Imperial Japan, which lasted from the Meiji Restoration of 1867 until the defeat of 1945, had none of these characteristics. Japan's geographical position, like those of peninsular Korea, Shantung, and Liaotung, was for all practical strategic purposes an extension of the Asian continent. In antiquity, Japanese warriors had been ferried across the narrow 78-mile-wide Tsushima Strait to battle the mainland Koreans and Chinese, and waterborne Japanese pirates had plundered the Asian coast. Otherwise, the Japanese had rarely ventured beyond these local waters.[3] To its ancient continental foe, China, Imperial Japan added Russia—first tsarist, then Soviet.

Since antiquity, the Japanese had been a farming and fishing people, ruled by landowning feudal aristocrat-warriors known as samurai who warred against each other. No merchant middle class existed. The samurai army leaders supported a semidivine emperor and dominated an authoritarian government, state religion, and agrarian economy. In short, Japan was a typical continental state. Upon the decision in 1868 to Westernize, the Japanese had to borrow heavily from Europe and the United States to obtain modern technology and training. And, in 1874, Japan created a state-owned merchant marine to transport the Imperial Army and to import necessary raw materials and manufactured goods. For 50 years, these ships were built in foreign yards.

But, if Japan was ever to become self-sufficient as an industrial state, it would have to expand beyond its mineral-poor home islands to the adjacent Asian landmass. Hence, Japan required a first-class army—trained initially by German military advisers and led, as always, by the samurai generals. Politically, the Meiji constitution even gave the military independence from the civilian ministers, whom it could remove from office if it disapproved of government policy. The army therefore enjoyed extraordinary power within the state.

Being strategically oriented to the continent of Asia, Japan had never entertained any serious maritime notions of having overseas colonies and possessions beyond the rimland of Asia. It had, therefore, in typical continental fashion, used a navy only to help defend the coastline and to transport and supply the army on the mainland. The Meiji regime saw no reason to alter this traditional role, and the Imperial Japanese Navy never escaped its subordinate role to the army. Consequently, it operated in support of the ground forces and thus never fought a naval battle of annihilation purely for the sake of annihilation. Furthermore, at no time during its 77-year existence did the Imperial Japanese Navy seriously try to protect its merchant marine or even attack enemy maritime commerce.[4]

Guarding the army's communications between the Japanese home islands and the Asian coast, however, posed a special problem. The coastal waters between Japan and the mainland—the Sea of Japan, Tsushima Strait, Yellow Sea, East China Sea, and South China Sea—had to be safeguarded from enemy naval attacks. And since both China and Russia were building modern fleets during the 1870s and 1880s, Japan followed suit, though only haltingly because of the army's resistance to this construction, culminating in a vicious attack on the navy by Japanese newspapers in 1893.[5]

Japan's admirals turned to the British and French for warships and training and sent several midshipmen to the U.S. Naval Academy.

From their Anglo-American teachers, they learned modern fleet tactics, and from the writings of Alfred Thayer Mahan they adopted the premise that the key to success at sea lay in the decisive big fleet battle. The Imperial Navy wedded this idea to its traditional mission of commanding its own local waters by planning to defeat any enemy fleet in an engagement close to home. The Japanese never did, however, until after the outbreak of World War II, plan to send their main battle fleet beyond the waters contiguous to Japan. Theirs was purely a defensive fleet—to defend the homeland and the army's seaborne communications.

Between 1874 and 1894, Japan sought to thwart the Chinese and Russians from extending their control over Asian coastal waters and lands but without resorting to war, for which Japan was not yet ready. A naval punitive expedition to Chinese Formosa in 1874 pointed out the need for a merchant marine, which Japan thereupon created. The same year, Russia established a naval base at Vladivostok and a decade later occupied Wonsan harbor in Korea until the displeased British forced the Russians to withdraw. Japan had great diplomatic success over this 20-year period by bargaining with the Western powers to allow it to occupy the several island groups nearest Japan's Pacific side—the Kuriles, Volcanos (principally Iwo Jima), Bonins, and Ryukyus (main island Okinawa).[6]

Concentrating on its major area of interest, the continent of Asia, Japan went to war against China over Korea in 1894–95. In this first modern Sino-Japanese war, the Imperial Navy supported the army's conquest of Manchuria and also crushed the Chinese fleet in one decisive fleet action in the Yalu River. Some Japanese leaders wanted to annex Korea outright, but the government feared this might lead to war with Russia. In the peace settlement, Japan was ceded Formosa.

The coveted prize was Port Arthur on the Liaotung Peninsula, control of which would have given Japan its first foothold on the continent of Asia. But the Western powers, at Russia's insistence, would not permit it. Worse, China then ceded Port Arthur to Russia for use as a naval base, whereupon the Russians lost no time in extending their control over the entire Liaotung Peninsula, northern Korea, and all of Manchuria by 1900.

The Continental State Upon the Sea

These aggressive moves by tsarist Russia were a direct invitation for Japan to build up its military forces, especially through purchasing warships from Britain. An expanding Russia had quickly supplanted the defeated China as the mortal enemy of Imperial Japan.

Although the Japanese Army planned for an ultimate land war against Russia for hegemony in Asia, the navy developed tactical plans to neutralize the Russian fleet based at Vladivostok and Port Arthur. These preparations were largely the work of Lieutenant Commander Saneyuki Akiyama of the Naval Staff College.[7] A student of Japanese as well as Western maritime history, Akiyama formulated a policy of attrition, whereby light fleet units—cruisers, destroyers, torpedo boats—would aggressively attack the approaching enemy fleet until its will was shattered and its strength so dissipated that the main battleships could engage on equal terms.

Such offensive tactics determined Japanese naval strategy as defensive: to wait for the enemy to come into the restricted Japanese waters. On the eve of the approaching war, Commander Akiyama reported on board the flagship *Mikasa* as staff planner to fleet commander Admiral Heihachiro Togo and his chief of staff, Captain Tomosaburo Kato.

In the Russo-Japanese War of 1904-05, Togo's Combined Fleet bottled up the Russian Pacific Fleet inside Port Arthur and repulsed its sortie. The resulting Japanese command of the Yellow Sea enabled the Imperial Army to be put ashore at Inchon in Korea and to drive across the Yalu into China with its seaward communications secure. Similarly uninterrupted was a second landing above Port Arthur on the Liaotung Peninsula. Waiting defensively for the Russian Baltic Fleet to arrive to save Port Arthur, Togo's fleet put Akiyama's battle plan into action. Togo caught the Russians in Tsushima Strait and virtually destroyed their fleet. The army then forced the surrender of Port Arthur, and peace was eagerly concluded by the exhausted belligerents.

The war had a lasting impact on Japan's continental strategy. To deter further Russian expansion in the Orient, the Western powers (principally the United States, Britain, and France) awarded Port Arthur to Japan and furthermore allowed Japan to dominate Korea.

In addition, Japan was given the Russian railroads transversing southern Manchuria, which the Japanese proceeded to develop and police with troops against the many bandits roaming the countryside. These new mainland possessions and protectorates gave Japan a strong defensive position against the Russians. And, in 1907, an Imperial Defense Policy declared Russia to be Japan's principal potential enemy. The Imperial Army thus remained the senior service, aimed at renewing the war on the continent.

As for the navy, because of the victory at Tsushima, Akiyama's strategy of attrition now became naval orthodoxy and remained the basis of Japanese naval doctrine for the next four decades—whoever the enemy. With the virtual elimination of Russia's fleet, the Imperial Navy had no difficulty in selecting a new primary foe: the U.S. Navy. The U.S. enforcement of the "Open Door" in China and occupation of the Philippine Islands as well as Guam, Wake, and the Hawaiian Islands since 1898 was climaxed in 1908 by the visit to Japan of the U.S. Atlantic battleships of the "Great White Fleet." The United States was viewed as a contender for hegemony in the Far East.

Captain Akiyama and his Naval Staff College colleagues applied their "whittling down" strategy in war games against the U.S. fleet in the Pacific. And, during the early months of World War I in 1914, Japan seized the German islands of the Central Pacific—the Palaus, Marianas, Carolines, and Marshalls—thus extending its buffer of islands into the open ocean. However, none of these islands were regarded as important enough to be fortified. Three years later, hypothetical Japanese war plans called for the navy's seizure of U.S. Manila and Guam in anticipation of a U.S. advance westward to invade Japan. "The Navy will try," said the plan, "to reduce gradually the force of enemy fleet units in transit and destroy them totally with our capital fleet units" in one decisive Tsushima-type battle.[8]

With Russia fighting in Central Europe and prostrated by its revolution, Japan extended its influence over the East Asian landmass by occupying German Tsingtao in China (1914), forcing "21 Demands" on China for broad political concessions (1915), and, with 70,000 troops, intervening in the Russian Far East, resulting in an undeclared war against the Bolsheviks (1918-22). Such aggressiveness aroused the United States to a naval arms race with Japan (1918-21), whose war planners promptly dropped the Soviet Union as the primary enemy in favor of the United States.[9]

By 1921, therefore, the Imperial Japanese Navy had achieved substantial political power alongside the army. Both the army and the navy faced the task of solving the problem of fighting and defeating the new major enemy. Neither the navy nor the Japanese people, however, wanted a war with the United States. Thus, the admirals attended the Washington Naval Conference late in that year, seeking ways to defuse the arms race without sacrificing necessary fleet strength for a potential conflict in the Pacific. The head of the Japanese delegation was Admiral Kato, Togo's chief of staff at Tsushima, who regarded the Imperial Navy as a deterrent force and who, despite the objections of the Navy General staff, convinced his government to accept the tonnage limitation of Japanese capital ships at 60 percent that of the U.S. Navy. This limit was renewed at the London conference in 1930.[10]

Since the navy considered 70 percent to be the absolute minimum of battleship strength necessary for meeting the U.S. foe in battle, Kato and his successors during the 1920s and 1930s turned to the late Akiyama's original strategy of attrition. Again, the key was tactics, but using new technology. Submarines, advocated Admiral Nobumasa Suetsugu, commander of the Combined Fleet from 1933 to 1937, would follow a "strategy of interceptive operations" to reduce the advancing U.S. fleet. Widely separated aircraft carriers would similarly throw their bombing and torpedo planes at U.S. ships to inflict more ship losses. Then, when the decisive battleship duel ensued, the Japanese would exploit night tactics, especially with their torpedo-launching destroyers, which had better speed and firepower than their U.S. counterparts. Such technological and tactical advantages would give the Combined Fleet the edge necessary to win.[11]

Japanese naval strategy thus remained as defensive as before, with the projected decisive battle to occur somewhere behind the line of the Bonins and Marianas—in the Western Pacific waters adjacent to Japan and the Asian littoral. And, as in the past, absolutely no thought was given to commerce warfare, either in protecting Japan's new privately operated merchant fleet or in attacking the enemy's. Suetsugu's submarines were to only attack warships.[12]

Just as the navy's outlook did not change, neither did the army's. The generals viewed with alarm the Soviet annexation of Mongolia in 1924, but it was the fanatical colonels of the Kwantung Army guarding Japan's rail lines in Manchuria who planned to do something about it. Several of these passionately anticommunist younger officers assassinated the local Chinese warlord in 1928 and three years later seized power over the entire province. Their goal was to make Manchuria a prosperous source of raw materials, an outlet for an overpopulated Japan, and thus a firm buffer against the continental Soviet Union. The government in Tokyo, unable to control these zealots, formally established the puppet state of Manchukuo to still discontent within the army. Henceforth, however, two army reform groups emerged—one content to beef up Manchukuo, the other aiming to control all of China. But both agreed on the same objective: to deter and defeat Soviet Russia on the continent.[13]

General and junior officers alike were restive with the simultaneous onset of the Great Depression and the perpetuation of Japan's inferior status in the London Treaty of 1930. Young army and navy zealots responded with several successful political assassinations and abortive coups between 1930 and 1933, which gave heart to the senior opponents of the London Treaty to be more aggressive. The generals and so-called "fleet faction" of anti-Western admirals caused the government to withdraw Japan from the League of Nations and not to renew the arms limitation agreements as of the end of 1936. Then, early that year, several young pro-Manchuria, anti-China expansion army officers staged an abortive coup in Tokyo. When it failed, many generals pressed the government for further expansion into China as a means of meeting the Soviet menace—in spite of the fact that a new navy-dominated Imperial Defense Policy edict that year reaffirmed the United States as Japan's main enemy.[14]

Internal events in China fed the fears of Japan's army officers. During 1936-37, the festering civil war between the ruling Chinese Nationalists and the insurgent Chinese Communists appeared to be nearing a truce based on the common Japanese foe. Dreading a Sino-Soviet alliance from such a rapprochement, radical Japanese Kwantung Army commanders in Manchuria launched raids into North China, culminating in a clash with Chinese troops at the Marco Polo bridge on July 7, 1937—two days after the signing of the official Nationalist-Communist Chinese pact against Japan. Ignoring the Army General Staff in Tokyo, which viewed an entanglement in China as drawing strength away from the Soviet enemy, the Kwantung Army engineered the second (and, until 1941, undeclared) Sino-Japanese War. Its goal was to conquer all China in a matter of months and then to turn against the Soviet Union. These warhawks drove the Chinese forces back, massacred between 200,000 and 300,000 civilians at Nanking, and incited all of China to unite against Japan.[15]

The war quickly became a quagmire, sinking the Japanese Army in a long, dirty ground war on the continent of Asia. More reasonable minds in the army could not extract their forces once a quick victory proved elusive. Worse, in March 1938, total national mobilization became necessary for Japan to fight effectively in China and to deter the unfriendly Soviet Union, and perhaps the United States and Great Britain as well. As a result, the civilians surrendered control of the government to the army, whose fears of the Soviet Union were borne out during the summers of 1938 and 1939 when Red Army and air force units severely punished Japanese forces in heavy fighting along the Manchurian-Mongolian borders. In July 1940, the army-led government of Japan declared its intention to create a Greater East Asia Co-Prosperity Sphere of China and Manchuria under Japanese direction. Such a unified continental empire would prevent further unfortunate Soviet "incidents."[16]

Concerned though the Soviet Union was, its alarm was minor compared with that of the United States. Self-appointed protector of China's national integrity ever since the "Open Door" policy at the turn of the century, the United States instituted gradual diplomatic and economic pressure on Japan to force it to withdraw from China. This took the form of import embargoes of vital war materials to Japan, covert military aid to China, and general rearmament for a possible war in the Pacific. Between the summers of 1940 and 1941, U.S.-Japanese relations worsened as the Imperial Army tried to decide on which potential enemy to face—the continental Soviet Union or maritime United States—while fighting the unpopular war in China.

To choke off Anglo-American aid to China over the Burma Road, Japan in September 1940 occupied northern Indochina (northern Vietnam), the impotent colony of France, which had recently capitulated to Adolf Hitler's German juggernaut. That same month, Japan formally allied itself with the Nazis, then besieging the British Isles by air and sea attack. English-speaking peoples viewed this alliance as another slap in the face. The Soviet question continued to plague Japan's generals until the diplomats eliminated the concern by signing a nonaggression pact with the Soviet Union in April 1941. Though heavy Japanese troop concentrations remained in Manchuria to watch the Soviets, Imperial Army pressure on China increased with the Japanese occupation of southern Indochina (southern Vietnam) in June. At this time, Germany invaded the Soviet Union, further easing Japan's fears of possible Soviet aggression in the Far East.

But Japan's seizure of Indochina caused the United States to freeze all Japanese assets and halt all oil exports to Japan—moves imitated immediately by Britain and the Netherlands—which had the effect of cutting off further fuel supplies to the Japanese war machine.

Negotiations between Japan and the United States resolved nothing, because of the Imperial Army's fear that a Japanese withdrawal from China, as the United States insisted, would invite Soviet aggression into China and Korea and ultimately mean Soviet control over the continent of Asia. Led by General Hideki Tojo, who became premier during the autumn of 1941, the army demanded that Japan conquer the Dutch East Indies and British Malaya, because their oil, rubber, and foodstuffs were essential for victory in China. After that southern offensive, the army planned to turn north again and attack the Soviet Union on the mainland, initiating the long-anticipated war on that front.[17]

Unfortunately for Japan, its southern drive required the seizure of the U.S. Philippines to cover the army's seaborne communications to the Indies and Malaya. In addition, the full energies of the Imperial Navy would be required for the traditional role of transporting, supporting, and covering the invasion forces.

The navy had never deviated from its defensive strategy, even as war with the United States became inevitable. It had been supporting the army in China, had seized offshore Hainan and the Spratly Islands in 1938-39, and had created a new fleet in the Kuriles to watch the

Soviets. But it made no serious attempt to fortify the Central Pacific islands against the United States until 1940.[18] From the airfields in these islands, long-range naval planes—especially the "Betty"— would join the submarines and carrier planes in whittling down the oncoming U.S. battle fleet. Although the admirals opposed the German alliance, the "fleet faction" of hawkish flag officers eagerly anticipated a war with the United States and gradually took control of the navy. Believing their own propaganda of moral superiority, they trusted their new *Yamato*-class superbattleships to destroy the U.S. Pacific Fleet when it arrived west of the Marianas, after having been reduced in the campaign of attrition.

A handful of senior admirals, however, remained opposed both to war with the United States and to the traditional decisive battleship action in the Western Pacific. Foremost among them was Admiral Isoroku Yamamoto, appointed to command of the Combined Fleet in 1939.[19] Yamamoto believed that the aircraft carrier would be the weapon to decide future naval battles, and the carriers must be used offensively, not in a passive defensive role. Upon taking command, therefore, he moved the area of the decisive battle from the Marianas and Bonins far eastward to the vicinity of the Marshall Islands.

Then, during the spring 1940 fleet maneuvers, successful carrier torpedo plane attacks so impressed Yamamoto that he seized upon the idea of possibly launching a surprise carrier air attack on the U.S. fleet anchored at Pearl Harbor to reduce its strength in one fell swoop. By July, he had extended the area of the decisive battle all the way eastward to include the Hawaiian Islands.[20]

If war with the United States must come, Yamamoto argued in a memo to the Navy Minister in January 1941, initial success depended on the neutralization of the enemy fleet in a surprise raid on its Pearl Harbor base by carrier planes, followed by a submarine blockade of Pearl. Most of the battleships would not participate in this hit-and-run affair. Surprise was essential. If, however, the U.S. fleet sortied for battle prior to the attack, Yamamoto insisted that the Combined Fleet, carriers as well as the battleships, must remain in the western Pacific for the battle as traditionally planned. So, instead of being the decisive naval engagement, the Pearl Harbor raid was designed as "a graduated measure based on the doctrine of offensive tactics in a defensive strategy in an effort to permit the endurance of a protracted war."[21]

Neither the army nor his navy superiors supported Yamamoto's scheme when they learned of it. Only by forceful arguments and by threatening to resign was Yamamoto able to gain acceptance for the "Hawaii operation." With the elimination of the U.S. carriers and battleships at their anchorage, the enemy's fleet could not challenge the invasion of Southeast Asia. In fact, if successful, the Pearl Harbor attack would virtually eliminate the U.S. Pacific Fleet, and the army could triumph in the "protracted war" in China. During the autumn of 1941, Hitler's drive into the Soviet Union promised victory there during 1942, which would enable the Japanese Army to grab the Soviet Union's Eastern provinces. With the Soviet Union defeated and the U.S. Navy crippled in the Pacific, Japan would realize its Greater East Asia Co-Prosperity Sphere on the continent.[22]

The 7 December attack only partially succeeded in whittling down the U.S. Pacific Fleet, but enough so to stop the surviving ships from heading west to prevent the fall of Southeast Asia to the main Japanese offensive. The U.S. carriers had been at sea during the air raid on Pearl; therefore Yamamoto returned with the bulk of the Combined Fleet to eliminate them in June 1942. Thanks to Allied codebreakers, the planned Japanese submarine blockade of Pearl that month failed to engage, much less whittle down, U.S. carrier strength, which in turn caught Yamamoto's four fleet carriers off guard and sank them all in the Battle of Midway.

And yet, Yamamoto succeeded by his strategy of attrition to forestall the U.S. advance across the Pacific. Off Guadalcanal, in the South Pacific during the remainder of 1942, his planes and submarines wore down U.S. carrier strength to one flattop, buying the Imperial Army time to fight its "protracted war" on the mainland of Asia. Not until June 1944 would the new U.S. Pacific Fleet reach the environs of the Marianas for the long-desired defensive decisive battle.

In that battle, the Battle of the Philippine Sea, however, the Combined Fleet was thoroughly defeated, because during the preceding two years the United States had turned the war of attrition—and industrial production—against Japan. Far worse, the Japanese Army had not only failed to win the war in China, but Hitler's promised victory over the Soviet Union had turned sour with the successful Soviet stand and counterattack. And Japan's generals continued to hold back many of their best troops in Manchuria in readiness for a Soviet attack, keeping them from fighting in China and the Pacific islands.

In July 1944, one month after the defeat of the Imperial Navy in the Philippine Sea, General Tojo's cabinet fell. But the government continued to be dominated by the army; it would not capitulate unless its main forces were defeated on the ground. The army's most effective enemy on the continent proved not to be the sprawling, corruption riddled, conventional Nationalist Chinese Army, but the guerrilla forces of the Communist Chinese. These so dissipated the energies and spirit of the Imperial Army in North China as to render it totally ineffective by 1945.

The defeat of large Japanese Army forces in the Philippines and on Okinawa that same year spelled the impotence of the senior service. Ironically, though, the Soviet juggernaut, which swept over the Imperial Army in Manchuria in August 1945, convinced the generals to surrender honorably.[23]

To the end, Imperial Japan had remained a continental power, dominated by its army. Naval analyst Alexander Kiralfy observed at the height of the Pacific war, "the navy had been primarily a protective shield for the homeland and for the military transport . . . , strictly a servicing unit for the Japanese armies."[24] Even when the navy and naval air forces had been rendered impotent, the army continued to fight until it, too, faced extinction by the invading Soviet Army and the U.S. atomic bombs.

In reviewing the history of the Japanese Empire, current U.S. defense planners and analysts should not allow Japan's many naval achievements to cloud the fact that Japan was first and foremost a continental nation, guided by continental objectives and a continental strategy. In our current strategic terminology, the Imperial Navy was purely a "sea denial" force, oriented to deny control of the Western Pacific by the U.S. fleet in order to allow the army to win the main war on the continent.

History and the Sea

At no time before or during the war did the Imperial Navy desire or attempt "sea control" of U.S. waters east of the Marshalls. The admirals never envisioned developing a Pearl Harbor of their own beyond the home islands. They developed the wartime Combined Fleet anchorage at Truk in the Carolines as little more than a makeshift expedient in July 1942, one month after Midway. Japan never seriously entertained the idea of seizing Hawaii or any other U.S. island to be a forward fleet base. Even had Midway Island been taken in 1942, it would have been used only as an airfield to keep Pearl Harbor neutralized. Similarly, the reason for the conquest of the Solomons and the projected seizure of Samoa in the South Pacific was to use their airfields to cut Allied sea lanes to Australia. These were all defensive strategic measures. Of course, any idea of a Japanese invasion of the U.S. West Coast was absurd.

Although no one would dare suggest there is a close parallel between the feudalistic-Oriental society of Imperial Japan and the current Soviet state, the potential strategic similarities warrant careful study and consideration. The strategic and naval experiences of Imperial Japan offer the closest example we have of a modern continental navy inflicting severe damage on a powerful maritime rival but, at the same time, being fatally handicapped by a dominant army. The experience of the Japanese Empire may thus teach us about the enemy we face today—a new continental empire with a first-class fleet of warships.

NOTES

1. Alexander Kiralfy, "Japanese Naval Strategy," Edward Mead Earle, ed., *Makers of Modern Strategy* (New York: Atheneum reprint, 1966), pp. 463–464.
2. Clark G. Reynolds, *Command of the Sea: The History and Strategy of Maritime Empires* (New York: William Morrow, 1974), pp. 12–15; Clark G. Reynolds, "The Sea in the Making of America," *Proceedings*, July 1976, pp. 36–51.
3. Arthur J. Marder, "From Jimmu Tenno to Perry: Sea Power in Early Japanese History," *The American Historical Review*, October 1945, pp. 1–34.
4. Kiralfy, pp. 470–474.
5. Fred T. Jane, *The Imperial Japanese Navy* (London, 1904), pp. 99–103.
6. John Curtis Perry, "Great Britain and the Imperial Japanese Navy, 1858–1905," Ph.D. dissertation, Harvard University, 1961, pp. 32–34, 49, 95, 111, 134, 153–157, 167, 192, 211.

7. Mark R. Peattie, "Akiyama Saneyuki and the Emergence of Modern Japanese Naval Doctrine," *Proceedings*, January 1977, pp. 60–69.
8. Sadao Seno, "A Chess Game with No Checkmate: Admiral Inoue and the Pacific War." *Naval War College Review*, January-February 1974, p. 28.
9. Captain Mitsuo Fuchida and Commander Masatake Okumiya, *Midway: The Battle That Doomed Japan* (Annapolis: U.S. Naval Institute, 1955), pp. 10–13.
10. Masanori Ito, *The End of the Imperial Japanese Navy* (New York: Macfadden, 1965), p. 9.
11. Rear Admiral Toshiyuki Yokoi, "Thoughts on Japan's Naval Defeat," *Proceedings*, October 1960, pp. 71–72; Stephen E. Pelz, *Race to Pearl Harbor* (Cambridge: Harvard University Press, 1974), pp. 28–29, 34–39; Ito, pp. 17–18; Clark G. Reynolds, *The Fast Carriers* (Melbourne, FL; Krieger, 1978), pp. 5–8; General Minoru Genda, "Tactical Planning in the Imperial Japanese Navy," *Naval War College Review*, October 1969, pp. 45–46. Genda's label for Japanese strategy is "diminution operation."
12. Yokoi, pp. 71 and 73; Stanley A. Wheeler, "The Lost Merchant Fleet of Japan," *Proceedings*, December 1956, pp. 1295-1296.
13. John Toland, *The Rising Sun: The Decline and Fall of the Japanese Empire, 1936-1945* (New York: Random House, 1970), pp. 7–11.
14. Toland, pp. 11ff.; Seno, p. 28; Pelz, pp. 10, 41–63, 174–175.
15. Toland, pp. 42ff.
16. Toland, pp.66, 74, 78–80; Lieutenant General Masatake Okumiya, "The Lessons of an Undeclared War," *Proceedings*, December 1972, pp. 26–28; Ito, p. 174.
17. Toland, pp. 139ff.; Pelz, p. 223.
18. Thomas Wilds, "How Japan Fortified the Mandated Islands," *Proceedings*, April 1955, pp. 401–407.
19. Another opponent, Admiral Shigeyoshi Inoue, in a paper in January 1941, very accurately predicted the course the war against the United States would follow. See Seno essay.
20. Ito, p. 26; Hiroyuki Agawa, *The Reluctant Admiral: Yamamoto and the Imperial Navy* (New York: Kodansha International, 1979), pp. 193–200.
21. Jun Tsunoda and Admiral Kazutomi Uchida, "The Pearl Harbor Attack: Admiral Yamamoto's Fundamental Concept," *Naval War College Review*, Fall 1978, pp. 83–88.
22. Fuchida and Okumiya, p. 55; Pelz, p. 223. The major thesis of Pelz's book is that the navy's race to build up arms was a major cause of war.
23. Raymond L. Garthoff, "Soviet Operations in the War with Japan—August 1945," *Proceedings*, May 1966, pp. 50–63.
24. Kiralfy, p. 483.

7

ERNEST J. KING AND AMERICAN MARITIME STRATEGY IN THE PACIFIC WAR

One of the few Allied leaders during World War II to appreciate the continental character of the Japanese Empire was Admiral Ernest J. King, as Commander in Chief United States Fleet and Chief of Naval Operations the wartime head of the U.S. Navy and counterpart of Army Chief of Staff General George C. Marshall. Though less well known to the public than his Pacific theater and fleet commanders Nimitz, Halsey, and Spruance, King was by any measure the leading maritime strategist in the war against both Japan and Germany. To use the title of Thomas B. Buell's 1980 biography, he was the premier "master of sea power."

Virtually alone, King had a comprehensive appreciation of the strategy required for achieving victory in the Pacific: the three-pronged island offensive to destroy Japanese naval power, coupled with the defeat of the formidable Japanese armies on the continent of

152

Ernest J. King and American Maritime Strategy

Asia by the allied Chinese and Russian armies—a strategy generally neglected or minimized by historians of the Pacific war.

This chapter was originally presented as a paper, "Admiral Ernest J. King and the Strategy for Victory in the Pacific War," at a joint meeting of the American Historical Association and the American Committee on the History of the Second World War in 1975 and was published in the Winter 1976 issue of the *Naval War College Review*.

American naval policy and doctrine from 1900 to World War II was oriented almost exclusively to the Pacific and Japan (save for World War I), an orientation thoroughly adhered to by Ernest J. King at least from the time he earned his aviator's wings in 1926. Unlike his interwar contemporaries in the Army who were either apathetic or pessimistic about a war in the Pacific,[1] King and the Navy in fleet maneuvers and theoretical studies fashioned a naval strategy designed to defeat Japan, the ORANGE enemy. King himself therefore developed a consistency of thought and singlemindedness of purpose which during World War II his peers and detractors alike found maddening.

In the fleet problems of 1931 and 1932, as captain of the aircraft carrier *Lexington*, King fought mock naval battles in the waters of the Galapagos and Hawaiian islands, respectively. In 1932–1933 he war-gamed a cross-ocean offensive against Japan as a senior student at the Naval War College. While Chief of the Bureau of Aeronautics, 1933–1936, as a rear admiral, King was Fleet Commander Adm. Joseph Mason Reeves' choice to command the U.S. Fleet in the Pacific should war come. In the rank of vice admiral, first as Patrol Plane Commander then Carrier Commander in the Pacific during 1937–1939, King toured American island bases, participated in two fleet problems involving attacks on Pearl Harbor, and devised carrier tactics. And while a member of the General Board, 1939–1941, Admiral King studied Philippine defenses among many other Pacific-related problems.[2]

Reprinted by permission, *Naval War College Review*.

It is no wonder then that, as wartime Commander in Chief, U.S. Fleet, King should finally apply his—and the Navy's—vast preparations for war in the Pacific. As Michael Howard has observed,

> That Pacific operations should occupy the principal place in his mind and heart was inevitable. On his shoulders rested the ultimate responsibility for the conduct of a war unprecedented in complexity and scope against an adversary whose skill and ferocity had astounded the world . . .; a war, moreover, to which America's allies could make only a marginal contribution.[3]

As a global maritime strategist and naval leader, Admiral King had no equal in the United States, British, or any other navy during World War II, which helps to explain his supreme self-confidence and dogged personality. Indeed, no admiral in American history had ever been faced with such Herculean tasks before King, forcing the historian to look elsewhere for a comparable figure, namely to Britain's Admiral Lord John Fisher. As First Sea Lord, 1904–1910, Fisher fashioned the modern Royal Navy and returned to lead it early in World War I, all with experience, genius, and determination. Arthur Marder's description of "Jackie" Fisher could equally befit "Ernie" King:

> . . . too assertive in his likes and dislikes of others and [who] could brook no opposition to his plans. In his zeal for the efficiency of the Navy he was no respecter of persons. . . . [He was] indiscreet, harsh, abusive, revengeful and increasingly autocratic. In a word, he was a very hard person to get along with, if one did not happen to agree with him.[4]

The British, in fact, appreciated both sides of King's disposition, the professional and the rugged. Said Field Marshal Sir John Dill early in 1944, "King does not get any easier as time goes on. I am ashamed of a rather sneaking regard for him. . . ."[5] Admiral Cunningham, King's British counterpart as First Sea Lord, saw King as

> the right man in the right place, though one could hardly call him a good cooperator. . . . A man of immense capacity and ability, quite ruthless in his methods. . . . He was tough and liked to be considered

tough and at times became rude and overbearing. . . . Not content with fighting the enemy, he was usually fighting someone on his own side as well.[6]

As "the forceful and unchallenged professional head of the navy," John Ehrman has observed, the difficult King "brought to the Joint Chiefs of Staff a clarity and sharpness in argument which would otherwise have been lacking" and which thus complemented the more patient and calm General Marshall, his opposite number in the Army.[7]

Admiral King's strategic genius lay in his general appreciation of the global dimensions of World War II, namely, the need to speed up the war in Europe in order to enhance operations in the Pacific, and in particular the nature of Pacific geography and how a strategy of concentration would defeat Japan. This historical policy—in which the enemy would be defeated piecemeal—had been practiced or theorized by the maritime British from Pitt the Elder in the Seven Years' War to Corbett's classic book of 1911 and Liddell Hart's "indirect approach" of the 1920s. First, the navy would isolate the enemy's homeland by commerce warfare and blockade to wreck the economy, would fight naval engagements to destroy the enemy's fleet, and would seize or neutralize the enemy's overseas possessions and strongholds. Second, the maritime nation would support, supply, and encourage a major continental ally for its great manpower reservoir by which its armies could then directly defeat the enemy on land.[8]

For the first part of his strategy of concentration against Japan, King envisioned and then masterminded a Navy-led drive into the Japanese mandated islands, forcing the Japanese to battle and destruction and culminating in the area of the Luzon Strait from which the blockade of Japan could begin. He had foreseen such a drive, focused on the Central Pacific from Hawaii to the Philippines via the Marianas, in his 1932 Naval War College analysis, a position in complete accord with the Army-Navy War Plans ORANGE and later RAINBOW FIVE.[9] In March 1942 he projected the initial basic strategy over the next two years as the establishment of "strongpoints" to protect Allied communications to Australia and to attack those of

Japan via the New Hebrides, Solomon, and Bismarck island groups.[10] At the Allied strategic conferences between January and August 1943, he directed Allied Pacific strategic policy in establishing the Central Pacific route of the Gilberts, Marshalls, Carolines, and Marianas to the Philippines and the Chinese coast. King also favored the concurrent South-Southwest Pacific drive against Rabaul and New Guinea, and his preference to bypass Rabaul eventually prevailed; but this theater remained secondary in his thinking and in fact. In March 1944, Part One of King's strategy of concentration became JCS policy: the "first major objective in the war against Japan will be the vital Luzon-Formosa-China coast area."[11]

When this was achieved, King believed that Formosa first rather than the Philippines should be taken as a necessary prelude to landing in China, but this possibility was scuttled in September 1944 by the lack of available troops for such an operation (and probably also because of General MacArthur's pro-Philippines stance).[12] In any case, King believed that an invasion of Japan proper would be unnecessary and that the blockade (and aerial bombing) would suffice, strengthened by bases along the Chinese coast.[13] By this strategy, King's navy and attached ground and air forces destroyed Japan's defenses in the Pacific and thus isolated the homeland by the summer of 1945.

The other part of King's strategy of concentration, the defeat of Japan's major ground forces by Allied manpower, emerged in the form of China. Throughout the war King argued that

> the key to a successful attack upon the Japanese homeland was the geographical position and the manpower of China. . . . Just as Russia warranted support to drain off German strength, China had to be kept in the war so as to occupy on the mainland of Asia heavy Japanese land forces and some Japanese air forces. . . . Chinese manpower [was] the ultimate land force in defeating the Japanese on the *continent of Asia.* . . .[14]

To keep China in the war, King placed great emphasis on the projected but unfulfilled British operations into Burma (ANAKIM), even to the point of offering amphibious craft and aircraft from the

Ernest J. King and American Maritime Strategy

Central Pacific during 1943.[15] When this failed to mature for lack of British interest and resources, King shifted his attention to projected American operations against Japan from bases to be seized along the Chinese coast, which would also complement his blockade program and underscore America's political commitment to China.[16]

King welcomed the projected Russian entry into the war in the Pacific, especially when the Chinese armies retreated late in 1944, but he continued to base his hopes on China.[17] In any case, this aspect of King's strategy of concentration succeeded, though only partly due to his indirect influence, as the Nationalist Chinese Army tied down the Japanese forces in southern China, the Chinese Communist Army neutralized those Japanese troops in the north, and the Russian Army rolled over Japanese forces in Manchuria in August 1945.

In realizing this strategy for victory in the Pacific, however, King faced five major obstacles throughout the war which consumed a great deal of his time and energy and against which he therefore exhibited great impatience.

The first obstacle was the insistence of British and U.S. Army leaders for more resources than King thought necessary to defeat Hitler first at the serious expense of the separate war against Japan. He urged a more vigorous prosecution of the war against Germany and Italy so he could maintain relentless pressure on Japan. But believing the Army had always expected to direct American strategy in wartime, King argued at length over most Army proposals and disputed such decisions as the Sicily landings as "merely doing something just for the sake of doing something."[18] When the British seemed indecisive on ETO plans for 1942, King and Marshall bluffed them with a proposal to shift idle American troops from Europe to the Pacific, a threat not lost on General Eisenhower who thus kept his offensive going in late 1942 lest he lose men and equipment to the hungry Pacific theater. But King's deep respect for and close working relationship with General Marshall overcame this obstacle, with Marshall concentrating most of his attention on the ETO and China-Burma-India (CBI) theater and King on the Pacific.[19]

King's second obstacle was Douglas MacArthur, who resented King's and thus the Navy's strategic leadership in the Pacific. MacArthur, in the words of his biographer, "never fully comprehended the principles of modern naval warfare,"[20] leading King to insist that MacArthur never have full operational control over Navy vessels, from the summer of 1942 when the issue was carriers for air support to the spring of 1945 and amphibious shipping for the invasion of Japan. Again, thanks to Marshall's touch, King and MacArthur stayed at arm's length throughout the war; compromises were hammered out over command relationships; and King only lost out to MacArthur on the question of the Philippines over Formosa.[21]

King's third obstacle was internal, namely, the several factions and strong personalities within the Navy who influenced Pacific strategy. As Commander in Chief, U.S. Fleet, King exerted direct control over the Pacific element of this fleet and its commander, Admiral Nimitz, meeting with him frequently during the war to agree on strategy and policy and occasionally—because of his impatience—ordering Nimitz to handle the fleet in a certain way.[22] In strategy, King not only solicited Nimitz' views but those especially of two other Pacific admirals, Richmond Kelly Turner and Forrest Sherman.[23] In November 1944 King strengthened his direct participation by shifting his Atlantic Fleet Commander, Adm. Royal E. Ingersoll, to the Pacific as Commander Western Sea Frontier and Deputy Cominch-Deputy CNO-Deputy Cincpac.[24] Logistically, King personally set landing craft and aircraft production schedules for the Pacific as well as the ETO.[25] In the submarine war, King established submarine base areas and quotas as well as target priorities, and in fleet surface tactics he influenced carrier operating formations.[26]

In personnel, King established the policy of aviator chiefs of staff for nonaviator commanders and vice versa;[27] ordered reductions in the size of amphibious and submarine staffs;[28] "strenuously opposed" 'spot' promotions;[29] and severely punished officers he disliked or otherwise opposed. Among the more celebrated cases were Comdr. "Mort" Seligman, executive officer of the *Lexington*, who was summarily retired for inadvertently leaking vital code-breaking knowledge to the press after the Battle of the Coral Sea;[30] Capt. "Carl" Moore, Admiral Spruance's chief of staff, who was passed over for flag rank for apparently failing to measure up to King's standards;[31] and Admiral "Jack" Towers, Nimitz' deputy and air type commander, who King kept from going to sea out of resentment over Towers'

Ernest J. King and American Maritime Strategy

ambitions to gain control of Pacific Fleet operations for himself and the air admirals.[32] King simply, for better or for worse, controlled flag personnel and their assignments, leading more than one unhappy admiral to utter Gilbert and Sullivan's line: "Just stay at your desk and never go to sea, and you will be ruler of the King's Navee"—Ernie's of course.[33]

The fourth obstacle to King's Pacific strategy was both internal and political—Britain's insistence on participating in the main naval operations against Japan late in the war. Since King regarded the Pacific war as America's show, he resented any British attempt to dictate strategy, much less to participate importantly in the fighting. King accepted the fact that Britain would reassert its hegemony in postwar Burma and Malaya, and he hoped the British would confine their wartime operations to the CBI theater and Indian Ocean, but he was suspicious of their possible postwar designs on the Dutch East Indies. Operationally, the British had inadequate logistical doctrines and facilities for the far-ranging operations of the Pacific theater, and King felt no compunction to allow the new British Pacific Fleet to use American bases and thus become a drain on them and possibly also on operations.[34] In this battle, however, Admiral King failed utterly, for in September 1944, at the second Quebec conference, President Roosevelt informed Prime Minister Churchill that the British Fleet would participate in the final operations against Japan. In the words of General Ismay, "The British delegation heaved a sigh of relief, and the story went the rounds that Admiral King went into a swoon and had to be carried out. . . ."[35]

King's final obstacle, and the most important one, was, of course, the Imperial Japanese Navy, the defeat of which he superintended in his capacity as Cominch-CNO. The Central Pacific campaign, "in size and brilliance of execution," says Arthur Bryant, the "rival . . . of Trafalgar,"[36] had been developed for at least ten years in King's mind before he first brought it to his peers on the JCS and Combined Chiefs of Staff. With single-minded determination, he persistently kept up the Allied pressure on Japan and never lost sight of the prime objectives of (1) destroying Japanese lines of communication for the eventual blockade and (2) attacking via the Marianas and Luzon bottleneck, both of which led to the battles that destroyed the Japanese Navy, naval air forces, and merchant marine—the key prerequisites for the defeat of Japan.

If his personality sometimes got in the way of others, what does it matter? No sane person has ever said that war is humane, civil, or an arena for the mild of heart. Admiral King saw what had to be done, and he did it in the most forceful possible terms. He was the right man in the right place at the right time—and with the right ideas—to direct the maritime strategy of concentration for victory in the greatest naval war in history.

NOTES

1. Russell F. Weigley, *The American Way of War* (New York: Macmillan, 1973), pp. 246, 259; and Forrest C. Pogue, *George C. Marshall, Vol. I, Education of a General, 1880–1939* (New York: Viking, 1963), p. 347. For the U.S. Army's traditional lack of interest in the Pacific, see Richard D. Challener, *Admirals, Generals, and American Foreign Policy, 1898–1914* (Princeton, N.J.: Princeton University Press, 1973), pp. 23–26; Graham A. Cosmas, *An Army for Empire: The United States Army in the Spanish-American War* (Columbia: University of Missouri Press, 1971), pp. 80–82, 102–107, and 121ff.; and Clark G. Reynolds, "American Strategic History and Doctrines: A Reconsideration," *Military Affairs*, December 1975, n.40 and passim. Rear Adm. Raymond D. Tarbuck, USN, wartime naval liaison officer on General MacArthur's staff, succinctly summarized Army ignorance of naval operations thus: "It is surprising how little the Army officers at GHQ know about water. . . . They treat even the smallest stream as an obstacle, but naval minds think of it as a highway." Quoted in D. Clayton James, *The Years of MacArthur, Vol. II, 1941–1945* (Boston: Houghton Mifflin, 1975), pp. 358–359.
2. E. J. King and W. M. Whitehill, *Fleet Admiral King* (New York: Norton, 1952), pp. 220ff., 228ff., 234ff., 264–265, 274, 279ff., 286–287, 303, 437 and n.
3. Michael E. Howard, *Grand Strategy, Vol. IV: August 1942–September 1943* (London: H. M. Stationery Off., 1972), p. 243. King, alone among the top American and British wartime commanders so preoccupied with the Pacific, was amused by the "not strictly pertinent high-priced thought" of persons who concerned themselves with Pacific strategy only very late in the war. King and Whitehill, p. 598.
4. Arthur J. Marder, *The Anatomy of Sea Power: British Naval Policy, 1880–1905* (London: Putnam, 1941), p. 394. See also *Richard Hough, First Sea Lord: An Authorized Biography of Admiral Lord Fisher* (London: Allen & Unwin, 1969), pp. 76ff. King replaced Admiral Harold R. Stark as CNO early in 1942 because, in the words of Robert E. Sherwood, Stark "lacked the quickness and the

Ernest J. King and American Maritime Strategy

ruthlessness of decision required in wartime," whereas King "lacked neither." *Roosevelt and Hopkins: An Intimate History* (New York: Harper, 1948), p. 164.

5. "He has built up a great Navy but he does not trust us a yard. . . ." Dill to Field Marshal Sir Alan Brooke, 4 February 1944, quoted in Arthur Bryant, *Triumph in the West: A History of the War Years Based on the Diaries of Field-Marshal Lord Alanbrooke, Chief of Imperial General Staff* (Garden City, N.Y.: Doubleday, 1959), p. 106.

6. "He was offensive, and I told him what I thought of his method of advancing allied unity and amity. We parted friends." Cunningham of Hyndhope, *A Sailor's Odyssey*, vol. II (London: Hutchinson, 1961), p. 134.

7. John Ehrman, *Grand Strategy, Vol. VI: October 1944–August 1945* (London: H. M. Stationery Off., 1956), p. 343. E. B. Potter has observed in a recent exchange on King in the *United States Naval Institute Proceedings*: "He really was a genius, respected for his ability, but genuinely loved by few." April 1975, pp. 75–77, for this and others opinions, notably Adm. J. R. Tate's denunciation of King. (One can only surmise the reaction of King, an inveterate ladies man himself, to then Captain Tate's impregnating of a famous Russian motion picture actress in 1945 in an incident recently publicized with the visit of the offspring, a Russian actress in her own right, with Tate in this country.) See also Harry Sanders, "King of the Oceans," *United States Naval Institute Proceedings*, August 1974, pp. 52–59, and J. J. Clark, "Navy Sundowner Par Excellence," *United States Naval Institute Proceedings*, June 1971, pp. 54–59. Marshall's biographer describes King's "bleakness of manner and rudeness in debate" and that he was "extremely jealous of the interests of the Navy." Forrest C. Pogue, *George C. Marshall, Vol. II, Ordeal and Hope, 1939–1942* (New York: Viking, 1966), p. 373. Pogue also recounts other British commanders' opinions of King in *George C. Marshall, Vol. III: Organizer of Victory, 1943–1945* (New York: Viking, 1973), p. 7. King opposed the appointment of a military adviser to the President as "detrimental to the interests of the Navy," meaning weakening King's direct access to the Commander in Chief, but the appointment of Admiral Leahy to that role made King change his mind. William D. Leahy, *I Was There* (New York: Whittlesey House, 1950), p. 96.

8. Sir Julian Corbett, *Some Principles of Maritime Strategy* (London: Longmans, Green, 1911), pp. 61–73; B. H. Liddell Hart, *Strategy*, 2d ed. (New York: Praeger, 1967), pp. 18–19; and Clark G. Reynolds, *Command of the Sea: The History and Strategy of Maritime Empires* (New York: Morrow, 1974), pp. 235–242, 290–291, 304–307, 413–416, 447–450. King never tired of urging his colleagues on the Allied CCS and American JCS to consider the three Axis Powers "as a whole"; without such global strategic planning, he said at Quebec (QUADRANT conference) in August 1943, operations in the Pacific were being hampered. King and Whitehill, pp. 483–484.

9. King and Whitehill, pp. 236–239, 242.

10. Ibid., p. 383, recounting King to Marshall, 2 March 1942. In November he tried to bypass the central and northern Solomons altogether in favor of the Admiralties, but his theater and area commanders Nimitz and Halsey would not agree to it. Louis Morton, *Strategy and Command: The First Two Years* (Washington: U.S. Dept. of the Army, Office of Military History, 1962), pp. 290–304, 370–371.

11. JCS to MacArthur and Nimitz, Serial 022941, 2 March 1944, quoted in George C. Dyer, *The Amphibians Came to Conquer: The Story of Admiral Richmond*

Kelly Turner, 2 vols. (Washington: U.S. Dept. of the Navy, 1972), vol. II, p. 857, also pp. 613, 855; Morton, pp. 382–384, 437–438, 443; Howard, IV, pp. 447–449; Ehrman, V, p. 432; Maurice Matloff, *Strategic Planning for Coalition Warfare, 1943–1944* (Washington: U.S. Dept. of the Army, Office of the Chief of Military History, 1959), p. 191; Pogue, III, pp. 208, 251, 253, 255, 439; King and Whitehill, pp. 419, 438–439, 444, 481, 485, 532, 534. On King and the bypassing of the North Pacific, see Pogue, III, pp. 156–157; on his role in the decision to bypass Truk, see Clark G. Reynolds, *The Fast Carriers: The Forging of an Air Navy* (New York: McGraw-Hill, 1968), pp. 116–119, 141–143.

12. Robert R. Smith, "Luzon versus Formosa," in Kent Roberts Greenfield, ed., *Command Decisions* (Washington: U.S. Dept. of the Army, Office of Military History, 1960), pp. 467–468, 473; Matloff, pp. 485–498.

13. King and Whitehill, pp. 575, 605n; Reynolds, *Fast Carriers*, pp. 322–324, 351.

14. King and Whitehill, pp. 419–420, 362, and 541–542. Italics original. On 5 March 1942, King told President Roosevelt (quoted on pp. 384–385), "The chief sources of manpower for the United Nations are China, Russia, the U.S., and to less degree, the British Commonwealth." Also, pp. 436, 506, 534; and Matloff, p. 34. On his view of Russian manpower in Europe, see Howard, IV, pp. 253–255.

15. Pogue, III, p. 25; King and Whitehill, pp. 418–420, 430, 510.

16. King and Whitehill, pp. 432, 440, 524.

17. Ibid., pp. 430–431, 516, 524–525, 542, 562, 591, 606; Pogue, III, p. 158.

18. Stephen E. Ambrose, *The Supreme Commander: The War Years of General Dwight D. Eisenhower* (Garden City, N.Y.: Doubleday, 1970), p. 158, this being at Casablanca. "King's war is against the Japanese." Dill to Churchill, 15 July 1942, quoted in Winston S. Churchill, *The Second World War, Vol. IV: The Hinge of Fate* (Boston: Little, Brown, 1950), p. 439.

19. Ambrose, p. 542; Pogue, II, pp. 340, 346, 372–373, 384, 387; King and Whitehill, pp. 367n, 390, 422, 442.

20. James, II, 359.

21. Pogue, II, pp. 255–256, 379–381; III, pp. 164–167, 169, 175, 206; Maurice Matloff and Edwin M. Snell, *Strategic Planning for Coalition Warfare, 1941–1942* (Washington: U.S. Dept. of the Army, Office of Military History, 1953), pp. 260–262; Jeter A. Isely and Philip A. Crowl, *The U.S. Marines and Amphibious War* (Princeton, N.J.: Princeton University Press, 1951), pp. 89–96; Richard M. Leighton and Robert W. Coakley, *Global Logistics and Strategy, 1943–1945* (Washington: U.S. Dept. of the Army, Office of Military History, 1968), pp. 579, 581, 590, 592–593, 604–606. MacArthur spread gossip through his British liaison officer, Lieutenant General Herbert Lumsden, early in 1944 "that King has finished serving his useful period, etc., etc." Brooke diary, 14 February 1944, quoted in Bryant, p. 106.

22. For example, King issued the bold Guadalcanal directive to Nimitz only 35 days in advance of the landings. "Only a great leader like Admiral King with great knowledge and great faith in his organization and the subordinates who were to lead their parts in it could have issued such a preparatory order." Dyer, I, p. 259. King sometimes pressed Nimitz and Halsey to act more aggressively, as in February 1942 throughout the Pacific, August 1943 in the South Pacific, in January 1944 in the Marshalls, and in July 1944 in the Marianas. Dyer, I, pp. 244–245; King to Halsey, 3 August 1943, personal correspondence file, Classified

163

Ernest J. King and American Maritime Strategy

Operational Archives, Division of Naval History; Thomas B. Buell, *The Quiet Warrior: A Biography of Admiral Raymond A. Spruance* (Boston: Little, Brown, 1974), p. 212; and Dyer, II, pp. 827, 932. One of King's strongest rebukes of Nimitz was his post-Gilberts demand for greater details to be furnished King on amphibious operations. Gen. Holland M. Smith and Percy Finch, *Coral and Brass* (New York: Scribner, 1949), pp. 138–139. King would not yield to Army pressure to split Nimitz' dual Cincpac-Cincpoa command for logistical reasons during 1943 partly because it would have violated King's direct control as Cominch over Cincpac. Morton, p. 478; Leighton and Coakley, pp. 444–448. Because of King's right to shift his flag as Cominch to sea, Under Secretary of the Navy James V. Forrestal in January 1944 tried, unsuccessfully, to get King out of Washington by having him assume direct command in the Pacific. Robert G. Albion and Robert H. Connery, *Forrestal and the Navy* (New York: Columbia University Press, 1962), pp. 125–126.

23. Dyer, II, p. 1143. In debates on strategy between King's and Nimitz' staffs in Washington, King "was pretty ruthless, but he listened to Forrest Sherman and he was inclined to follow his advice. It was my impression that King wasn't listening much of the time when others were talking. He always leaned forward attentively when Forrest Sherman was talking." Capt. G. Willing Pepper, USNR (Ret.) of the wartime Cincpac staff, to the writer, 16 August 1966.

24. King and Whitehill, p. 581; Samuel Eliot Morison, *Victory in the Pacific* (Boston: Little, Brown, 1960), p. 157.

25. Howard, IV, p. 430; Reynolds, *Fast Carriers*, pp. 229, 324. The inability of Army Service Forces to accommodate the long-jump Navy offensive in the Pacific helped to frustrate King's plans for Formosa in 1944. Leighton and Coakley, pp. 407–408, 413–415; Morton, pp. 492–493.

26. Clay Blair, Jr., *Silent Victory: The U.S. Submarine War Against Japan* (Philadelphia: Lippincott, 1975), pp. 201–203, 306, 361, 474–475, 577, in which King is criticized for basing subs in Australia in 1942, for his target priorities of carriers and battleships over merchantmen, and for his unimaginative use of the submarines. Reynolds, *Fast Carriers*, pp. 29–30, 228, 233, and passim.

27. Reynolds, *Fast Carriers*, pp. 120, 233–234.

28. Dyer, I, pp. 211, 216; Blair, p. 813.

29. Pepper to writer, 16 August 1966, recounting the bitter but victorious battle of Admiral Towers against King for Pepper's spot promotion to lieutenant commander on the ComAirPac staff.

30. Recounted in Blair, pp. 258–260.

31. Buell, pp. 220–251. King apparently criticized Spruance himself for being too cautious after the Battle of Midway, but King's respect for Spruance's overall ability no doubt saved the latter, p. 158.

32. Reynolds, *Fast Carriers*, pp. 68–69, 233; King to Nimitz, 12 August 1943, personal correspondence, Classified Operational Archives, Division of Naval History. Also, King failed to have some of his own cronies break into high command in the Pacific, like Adm. Alva D. Bernhard. Reynolds, *Fast Carriers*, p. 91. See p. 385 for my favorable conclusion over King's role in creating the air Navy.

33. A variation of W. S. Gilbert's famous act I statement from "H.M.S. Pinafore," quoted by Rear Adm. Joseph C. Cronin, USN (Ret.), in reference to Adm. J. F. Shafroth, to the writer, 8 September 1965.

34. Howard, IV, pp. 243–253, 278, 283, 563–564; Ehrman, V, pp. 432–433, 451, 522–523; Pogue, III, p. 28; King and Whitehill, pp. 511, 611; Matloff, p. 495; Reynolds, *Fast Carriers*, p. 302. In 1941–1942, King also showed great distrust of the Dutch in the war against Japan. King and Whitehill, p. 368n; Blair, p. 175.
35. "But this was an exaggeration." [Ismay Hastings], *The Memoirs of General Lord Ismay* (New York: Viking, 1960), p. 374.
36. Bryant, p. 27.

ADDENDUM

On 8 February 1944 King wrote to Nimitz, "You, [C. H.] McMorris, and [Forrest] Sherman have well expressed the ultimate necessity of closing in on the Japanese on the CHINA-KOREA side." King believed that the United States needed China as a base and "Chinese manpower to secure and maintain that base," also Chinese ports. Two days later Nimitz instructed Vice Admiral Frank Jack Fletcher, Commander Northwestern Sea Frontier, to study the possibility, however remote, of Russia coming in.

Over the spring the complicated problems of supplying both American and Russian forces in Siberia were set forth in JPS 467 (Revised) of 8 June 1944, "a good study" Nimitz told Fletcher, then Commander North Pacific Force and Ocean Area, on the 22nd of that month. "Our strategic concept still includes the possibility of a northern assault in 1945. Russia may become a belligerent but there seems to be assurance to that effect." Nimitz also observed, "Bombing the Japanese homeland may not be decisive but it is another means to that end. Every blow counts. . . ."

From Chester W. Nimitz correspondence, Series XIII, Number 7, Naval Historical Center.

8

DOUGLAS MACARTHUR
AS MARITIME STRATEGIST

Like Ernie King, General Douglas MacArthur was oriented to the Pacific and Asia rather than to the Atlantic and Europe, in contrast to his fellow U.S. Army officers. Indeed, the Pacific so dominated his strategic thinking that he deeply resented the Navy's dominant role in that ocean, a strategic preserve of the Navy before and during the Pacific war. Furthermore, MacArthur's unabashed anti-Navy, anti-Marine Corps rhetoric made him few friends in the Naval Service during and after World War II. This stance has tended to cloud historians' appreciation of the reality that MacArthur was a maritime strategist of the highest order who shared King's understanding of the dual necessity for victory on the continent of Asia as well as in the Pacific islands.

What is more, MacArthur assumed the Cold War role of maritime strategist by reconstructing Japan to be a bulwark of American defenses in the Pacific and by directing United Nations forces along maritime strategic lines during the first year of the Korean War. In so doing, he underscored his World War II genius in the application of maritime strategy in the broadest sense. This chapter was originally given as a paper before the Pacific Coast Branch of the American Historical Association in 1979 and published in the March–April 1980 issue of the *Naval War College Review*.

"I Have Returned"°

When Doug MacArthur at last went back,
"I have returned,"
He followed the fox-hunting, pig-boats's track,
"I have returned."
Oh, the carrier planes were overhead
And the battleship turrets spouted lead
So the general could go ashore, it's said, singing
"I have returned."

Cincpoa°° divisions were at his side,
"I have returned."
And Phibspacfor† provided the ride
"I have returned."
Oh, the subs went up to the sea of Japan,
And the carriers ranged from Saipan to Bataan,
So the general could land, according to plan, singing
"I have returned."

MacArthur petitioned God by prayer,
"I have returned."
But God decided He couldn't be there,
"I have returned."
But to help . . . God went to the utmost limits
For God sent Kinkaid, Halsey, and Nimitz
And MacArthur went along to kibitz . . . singing
"I HAVE RETURNED."

Such is the image, portrayed by an unknown wartime wag, that
the U.S. Navy had of "Dugout Doug" during the Pacific war—a
publicity hound and strutting blowhard who reckoned himself second
only to the Almighty, a landlubbing soldier with no real appreciation
of the Navy and its Marine Corps; guilty, more or less, on all counts.
But these superficial trappings must be peeled away from the aura

---

°Anonymous. Sung to the tune "When Johnny Comes Marching Home" by members of Air
Group Three in *Yorktown* (CV-10), December 1944.
°°Commander in Chief, Pacific Ocean Areas.
†Amphibious Forces Pacific Fleet.
Reprinted by permission, *Naval War College Review*.

surrounding Douglas MacArthur if we are to discover the essence of his strategic thinking, his *maritime* strategic thinking.

Maritime strategy is not naval strategy. Naval strategy may be defined as the employment of Navy forces to a specific end. Maritime strategy has a much broader scope: the combined use of all arms— Army, Navy and Air Forces—in seaborne operations. Historically, it has meant the "indirect approach" of maritime nations against their continental enemies by first winning command of the sea, destroying the enemy's seaborne commerce, conducting a naval blockade, and making amphibious assaults on outlying enemy territory and colonies—all designed to completely isolate the enemy's homeland, the classic formula propounded by Sir Julian Corbett in 1911.[1]

As employed by Britain in the Seven Years' War, in the Napoleonic wars and, unsuccessfully, in World War I, this indirect approach had meant a strategy of concentration: defeating or neutralizing France's or Germany's naval and maritime power and supporting a major allied army with money, arms, equipment and encouragement to decide the final issue on the continent—Prussia in the 1750s and Russia in 1812 and 1914. It meant that the maritime power could not afford to make massive commitments of her own ground forces on the continent, as Britain had done on the Western front during World War I, at least not until the enemy's army was in irreversible strategic retreat. Then an expeditionary force, like Wellington's, would land to help the ally's army administer the final blow.

Strategically, Britain has been an island nation simply incapable of shouldering the sheer immensity of both a large-scale ground war on the continental landmasses and a naval war. So too has this been a strategic limitation of the United States, whose maritime strategists from Alfred Thayer Mahan at the turn of the century to Adm. Ernest J. King in World War II worked diligently to hammer out a maritime strategy for this country.

To their names, one must add Gen. Douglas MacArthur. Though a typical U.S. Army officer of the first four decades of the century in his total concern with "occupation and pacification operations" (except for World War I), he matured into a major maritime strategist during World War II and the Korean war, alone among senior American generals to that time to do so.[2]

History and the Sea

MacArthur's maritime field throughout his career was the Pacific. Not only did he grow up during the era of Mahan, but at the time that he entered West Point, 1899, his father, Gen. Arthur MacArthur, was fighting insurrectionists in the Philippines as was his brother, an ensign of the same name in the Navy. Upon graduation in 1903 MacArthur went directly to those islands for his first duty and thence to visit Japan, an experience, he later reminisced,

> without doubt the most important factor of preparation in my entire life. . . . The true historic significance and the sense of destiny that these lands of the western Pacific and Indian Ocean now assumed became part of me. They were to color and influence all the days of my life. . . . It was crystal clear to me that the future and, indeed, the very existence of America, were irrevocably entwined with Asia and its island outposts. . . .[3]

This American and MacArthur's destiny in the Pacific closely involved the general during the interwar period, with two tours of duty in the Philippines (1922–25, 1928–30) before his term as Army Chief of Staff, 1931–35, during which time Japan invested Manchuria to begin its aggression in the Pacific. MacArthur then returned to the Philippines as Military Adviser and field marshal in 1935, being recalled to active duty in 1941 as commander of U.S. military forces in the Far East. Throughout these twenty years, during which no Army generals were really contemplating strategy, and for the initial unhappy days of World War II in the Pacific, MacArthur was preoccupied with the defense of the Philippines and merely had to accept War Plan Orange which called for a Central Pacific naval relief of the islands in wartime. His involvement there, however, made the Philippine Islands into the keystone of MacArthur's future maritime strategy for the eventual defeat of Japan.

Upon the fall of the Philippines and all Southeast Asia to the Japanese during the winter of 1942, General MacArthur turned his attention to the strategy whereby the United States might defeat Japan—and liberate the Philippines in the process. From his first thoughts on the subject, he embraced a maritime strategy. That is, the United States must defeat Japan not by a direct invasion of Japan proper or even by landings on the continent of Asia, in China, but by the indirect approach of whittling down Japanese power in battles for the seizure of the Pacific islands.

Once Japan had spent her naval and air forces in futile attempts to defend the Solomons, New Guinea, Halmaheras, Philippines and Ryukyus against American landings, the home islands of the Japanese Empire could be blockaded and surrounded by superior U.S. naval and air forces. MacArthur himself would command the ground forces of these assaults. The armies of Nippon in China would be bypassed, tied down in the prolonged war against the Chinese. But under no circumstances must the United States attempt to pit its ground forces against the huge Japanese armies in China or Japan.

The frustrations of achieving this end were not in theory, but in the details of execution. Indeed, MacArthur's maritime strategy closely duplicated that of Admiral King and Gen. George C. Marshall of the Joint Chiefs of Staff. The principal opposing detail was the divided command into two main theaters—MacArthur's Southwest Pacific and Adm. Chester W. Nimitz' Central Pacific. From this division emerged two preferred routes to Japan: MacArthur wanting the single line of New Guinea-Philippines, King the dual approach of the Marshalls-Marianas-Formosa with New Guinea secondary and the Philippines to be bypassed.[4]

MacArthur's futile attempt to have the JCS subordinate Nimitz' Central Pacific Force to his own drive up the New Guinea coast during 1943 brought on the not unreasonable fear of the Navy that its new aircraft carriers would be expended guarding MacArthur's northern flank. MacArthur's ubiquitous communiques, which almost failed to recognize the Navy's contribution to the war, only confirmed Admiral King's worst fears—and cemented King's opposition to MacArthur.[5]

None of this rivalry, however, should detract from the fact that General MacArthur steadfastly maintained his commitment to a maritime strategy aimed at choking off Japan's interior sea lanes at the "Luzon bottleneck." Furthermore, in January 1944, just as the dual counteroffensive was in its initial stages, MacArthur found adherents to his program in two very influential admirals, John H. Towers, Nimitz' Deputy and air commander, and Forrest Sherman,

Nimitz' brilliant planner and another naval aviator. Both men argued for the Pacific drives to meet at the Philippines and convinced Nimitz to agree with them.[6]

Two months later, in March 1944, the JCS agreed on going into the southern Philippines (but with MacArthur still splitting command responsibilities with Nimitz); he was to aim for Luzon in the northern Philippines and Nimitz for Formosa, both in early 1945. Admirals King and Nimitz pressed for Formosa and the adjacent Chinese coast at Amoy over Luzon for the next six months. Both arguments were thus discussed at length in Hawaii in July 1944 when President Franklin Roosevelt met with MacArthur and Nimitz to review the alternatives. For political as well as military reasons, MacArthur strongly opposed the Formosa plan in favor of Luzon.[7]

A flaw in the Navy's strategic preference for Formosa-Amoy lay in the second part, Amoy. To land on the mainland of Asia as well as in heavily defended Formosa, merely to have naval base facilities and airfields for closing the blockade of Japan proper, flew in the face of sound maritime strategy. Amoy would have to be defended—not as a beachhead or bridgehead but as an enclave, a blockading station, against massive concentrations of still unbeaten Japanese regular Army forces. Amoy would then have to be defended by strong continental forces to prevent a possible Dunkirk and that meant compromising the maritime strategy and fighting continental battles away from Japan's strategic center—Japan itself. Everything for Formosa-Amoy depended upon the ability of the Nationalist Chinese armies to keep open U.S. airfields in China to support this enclave.

In September 1944 General Marshall turned the Army's position against Formosa-Amoy and for the easier target of Luzon. JCS Chairman Adm. William D. Leahy agreed, and so did Admiral Nimitz as the Japanese Army began to overrun the U.S. B-29 bases in China. Admiral King finally had to go along, so that by the time MacArthur's forces landed at Leyte in the Central Philippines in late October MacArthur's maritime strategy had prevailed.[8]

The next decision involved the final campaign to defeat Japan, and here MacArthur's clear maritime strategic thinking reached its zenith. General Marshall and the Army insisted on the invasion of Japan in the traditional continental style as at Normandy in June.

Although the Allied Combined Chiefs of Staff in December had suggested invasion might be unnecessary, the JCS nevertheless decided in July 1944 to plan for such invasion one or two years hence.

Admiral King and the Navy would have none of this. Japan must be sealed off and strangled into submission by air-sea blockade—a philosophy with which the Army Air Forces thoroughly agreed. But King and the Navy still wanted an Amoy-type landing now further up the Chinese coast near Shanghai at the Chusan Archipelago, along with Okinawa Island in the Ryukyus.[9]

As this debate developed, Adm. Forrest Sherman flew to MacArthur's advanced headquarters at Leyte early in November 1944 to iron out some operational details, but during an evening of conversation MacArthur spoke frankly with this man whose mind he obviously respected. (Quoting Sherman's notes in full,) MacArthur

> Expounded [his] theory of our naval and air superiority in the Pacific, and came out flatfootedly against fighting [the] Japanese Army. Particularly emphatic against invading Japan. Said his troops were keyed up to take the Philippines, but basically wanted to go home after that. Said the Fleet should be freed from amphibious war to exert its power to isolate Japan same for air. I emphasized [the] extent to which POA had done only such fighting on shore as was necessary to get bases. He emphasized [the] extent to which he had been able to advance and "cut off half a million men" "under the wing of the Fleet."

After that half million left behind on the bypassed islands, MacArthur preferred not to face the more than two million Japanese troops on the home islands or another two million-plus on the continent, in China, Manchuria, Korea and Indochina.[10]

At the very same meeting, however, the general "stressed several times the ultimate operation on the 'Plains of Tokyo.' This seeming contradiction may be explained by MacArthur's urgent desire to eliminate as many more Navy-run amphibious operations as possible and to tighten the noose around Japan proper. So while he could not thwart Iwo Jima and Okinawa, neither of which subsequent operations during the first half of 1945 impressed him, he had no use for the Chusan operation being sought by the Navy during the early months of 1945.[11]

"The plain of Tokyo" had become the Army's avowed goal, with a preliminary assault on Kyushu during the autumn of 1945, Operation OLYMPIC, and though General MacArthur preferred not to fight the huge Nipponese Army he could see very well that OLYMPIC was in the works and that he was the logical choice to command it. And if the invasion must occur he had the answer to minimizing his losses: to implement fully a maritime strategy of concentration.

Secretary of the Navy James Forrestal at Manila in February 1945 recorded MacArthur's strategy: He

> expressed the view that the help of the Chinese would be negligible. He felt that we should secure the commitment of the Russians to active and vigorous prosecution of a campaign against the Japanese in Manchukuo [i.e., Manchuria] of such proportions as to pin down a very large part of the Japanese army; that once this campaign was engaged we should then launch an attack on the home islands, giving, as he expressed it, the *coup de main* from the rear while substantial portions of the military power of Japan were engaged on the mainland of Asia. . . . He said he felt that our strength should be reserved for use in the Japanese mainland, on the plain of Tokyo, and that this could not be done without the assurance that the Japanese would be heavily engaged by the Russians in Manchuria.[12]

He told a fellow officer that same week, ". . . we should make every effort to get Russia into the Japanese war before we go into Japan. . . ."[13]

The classic maritime strategy: blockade by the maritime power at sea and from the air while a major allied army—Soviet Russia's—fights on the continent, with the maritime power's expeditionary force—led by MacArthur—landing in the enemy's homeland to administer the *coup de grace* from the rear. Such had been the strategy of Eisenhower at Normandy while the Russians drove the Germans back on the main Eastern Front. Such had been Wellington's strategy against the French in Spain as Napoleon retreated from Russia after 1812. And such was Admiral King's identical thinking on this war in the Pacific.[14]

Given command in May 1945 of the ground forces that would land in Japan, MacArthur pressed the JCS for promoting the Russian attack on Japan, unaware that the Russians had already promised it at the Yalta Conference early in the year. He wanted more than 60 Russian divisions to be used in China as against maybe 13 American divisions to be employed in OLYMPIC and 22 for CORONET, the attack on Tokyo and Honshu in early 1946. Rejecting the idea of bogging down American forces on the mainland of Asia, in April he told the JCS that the American invasion of Japan "would continue the offensive methods which have proven so successful in Pacific campaigns. . . ."[15]

In other words, now that the air-sea blockade had isolated Japan, and assuming that the Russians would pin down the Japanese armies on the mainland, MacArthur put his trust in the maritime-amphibious forces that had so impressed him ever since the landings at Hollandia, New Guinea and at Leyte.

His biographer D. Clayton James has observed that MacArthur "never fully comprehended the principles of modern naval warfare, especially the complexities and dangers inherent in operating fast carrier groups, the Navy's most potent striking force."[16] This had changed, however, by the summer of 1945. MacArthur's considerable powers of observation, somewhat at Hollandia but most profoundly at Leyte, had schooled him in naval operations and taught him to distinguish between effective and ineffective leadership of these naval forces he was learning to respect so much.

For example, within days of Adm. William F. Halsey's near debacle at Leyte Gulf in which Halsey had perilously uncovered MacArthur's beachhead to a Japanese surface fleet attack by taking Adm. Marc A. Mitscher's fast carriers to hit the Japanese decoy carriers, the general changed his theretofore high opinion of "Bull" Halsey. In talking to 7th Fleet Commander Adm. Thomas C. Kinkaid, according to Admiral Sherman, MacArthur "apparently . . . criticized Halsey bitterly for [his] tendency to leave his covering assignment, and questioned [the] accuracy of his claims of damage inflicted." Kinkaid however calmed down MacArthur enough for him to send a congratulatory note to Halsey for his victory at Leyte.[17]

When Admiral Sherman had visited MacArthur at Tacloban on Leyte in November, MacArthur heaped praise on the escort carriers and ships' gunfire but "felt that Halsey by rushing off to watch

Mitscher strike [the enemy] CVs had created a critical situation which had been retrieved by [the] magnificent work of [the] CVEs and their escorts, but [he] seemed to speak entirely without rancor."[18]

Contrasting the easy handling of Admiral Mitscher's 5th Fleet carriers at Hollandia in April 1944 with Halsey's 3rd at Leyte in October and November, MacArthur

> spoke highly of the Fifth Fleet and [of] Mitscher's work at Hollandia [and] said: "Off the record—give me the Fifth Fleet or any other than the Third." [He] referred disparingly [sic] to Halsey's "wise cracks" which were not helpful when he wanted information. . . . Said he never knew where Halsey was—neither did Kinkaid. Referred to him as "not a team player." All this when annoyed at lack of knowledge of Halsey's plans. . . .

Two days later MacArthur "commented on Halsey's failure to reply to his messages and his complete ignorance of Halsey's position. It is noteworthy that without [radio] intercepts he has no picture of the strategic position at sea."[19]

Significantly, for the anticipated invasion of Japan, 5th Fleet under Adm. Raymond A. Spruance would provide the assault ships and close support, expanding on its successes in the Marianas and at Iwo Jima and Okinawa. And whereas before these operations Mac-Arthur had been a notorious Marine-hater as well as ignorant of Marine and Navy close air support techniques, after them he fully accepted the superior techniques over Army Air of Navy and Marine Corps close support as well as the complete trustworthiness of the Navy-Marine assault team. For OLYMPIC, therefore, he enthusiastically included Marine Corps divisions and tactical air units and insisted upon Navy control of all close air support which would include 16 escort carriers, four with Marine air groups.[20]

As his plans matured, MacArthur warmed to the prospect of the final drive into the home islands, but he could not be assured that his maritime strategy would succeed completely until Russia attacked the huge Japanese army in China. When that attack came in early August, MacArthur was thrilled as he learned the details of the Russian advance. His strategy had been rounded out.[21]

Douglas MacArthur as Maritime Strategist

The end came within days of the Russian invasion of Manchuria and of the two atomic bombings, that final act of the air-sea blockade. A maritime strategy had defeated Japan, with the added bonus that MacArthur had never had to lead that last assault he had wanted to avoid all along. Later he described the American strategy for victory.

> By the time we had seized the Philippines, we were enabled to lay down a sea and Navy blockade so that the supplies for the maintenance of the Japanese armed forces ceased to reach Japan. . . . At least 3,000,000 of as fine ground troops as I have ever known . . . laid down their arms because they didn't have the materials to fight with . . . and the potential to gather them . . . where we would attack. . . . The [Allied] ground forces that were available in the Pacific were probably at no time more than one-third of the [Japanese] ground forces; but . . . when we disrupted their entire economic system, . . . they surrendered.[22]

Though preoccupied with reconstructing Japan in the five years after V-J Day, General MacArthur never deviated from his commitment to the maritime strategy which had worked for him in the Pacific war. He held fast while the cold war situation deteriorated in the Far East with the overthrow of Nationalist China by the Communist Chinese late in 1949 and several concurrent fears; the possibility that Soviet Russia might assist the Red Chinese against the Nationalists on Formosa; the threat of the Vietminh against the French in Indochina; the fighting of the Huks in the Philippines; and the North Korean menace to South Korea.

In February 1950 the JCS—led by its Chairman, Gen. Omar Bradley, but intellectually dominated by Adm. Forrest Sherman—journeyed to Japan to discuss American strategy in the Western Pacific with the Supreme Allied Commander there, MacArthur. To deter all Asian communists, the Chiefs now gave MacArthur control over all fleet units whenever they were in Japanese waters rather than just during emergencies, and reaffirmed the focus of U.S. naval forces to be Yokosuka in Japan and Subic Bay in the Philippines. MacArthur asked that amphibious units training with the Marines be diverted to the Army, but the JCS turned him down.[23] Ominous intelligence

projections followed the JCS visit, but MacArthur had no firm strategic plans for any conflict in the Far East.

The bedrock of MacArthur's Far East strategy, however, lay in
preserving the integrity of America's bases in the "littoral island
chain" that ringed the Asian mainland. He told visiting retired Adm.
Charles M. Cooke in April 1950 that the Philippines and Formosa
were the keys, but Formosa occupied his attention. Whereas we had
morally liberated the Philippines and remained pledged to their
defense, Formosa had become a purely military asset whose importance had to be impressed upon the U.S. government. MacArthur told
the JCS this in February, and on 14 June 1950 he sent a four-and-a-
half page typed single-spaced "Memorandum on Formosa" to them,
adding that Formosa could be a springboard for attack to the south, as
it had been for the Japanese in 1941–42.[24]

The North Korean attack across the 38th parallel within days of
this memo plunged MacArthur back into the role of active strategist—
a Pacific *maritime* strategist. As if the five years since he had been
planning Operation OLYMPIC had been but a minute in time, he
harkened back to his hopes for an allied Russian attack into Japanese
China by advising the JCS in July to unleash Nationalist Chinese naval
and air forces from Formosa for attacks on mainland amphibious
targets and airfields, where Soviet advisers and airplanes were in
evidence.

Meanwhile, he employed U.S. naval and air forces to isolate the
fighting in South Korea while his ground forces stabilized their retreat
at the Pusan perimeter. The prospect of Americans fighting the
ground war in Asia that he had managed to avoid in World War II,
especially now with only limited forces, did not appeal to him at all
and in August he confided to the Chief of Naval Operations, Sherman,
"that we must turn Korea over to Koreans as soon as [the] fighting is
over and not occupy it."[25]

The association of the two kindred minds of Douglas MacArthur
and Forrest Sherman was an intellectual one, stemming from their
few meetings during the conduct of the Pacific war but which now
reached full flower in the strategy over the Korean conflict. They
were both maritime strategists who had agreed on the strategy that
had defeated Japan and who now concurred in a dual goal in the new

struggle: (1) to retrieve Allied reverses in Korea, and (2) to keep the war from expanding into Chinese territory.

For the first problem, winning the battle in Korea, MacArthur conceived the ingenious scheme in July of landing an expeditionary force in the communist rear at Inchon, retaking Seoul and routing the North Korean army. The sheer audacity of the plan brought Sherman and Army Chief of Staff Gen. J. Lawton Collins out to Japan late in August to study it more closely. When they arrived MacArthur had nothing but praise for the Marines, Air Force and carrier airplanes holding the line at Pusan. And now he wanted more troops for the Inchon assault, he said, namely the Army's X Corps and the First Marine Division, for a miniature OLYMPIC. To the doubts of the JCS over the wisdom of the enterprise, said MacArthur:

> My confidence in the Navy is complete, and in fact I seem to have more confidence in the Navy than the Navy has in itself. The Navy's rich experience in staging the numerous amphibious landings under my command in the Pacific . . . leaves me with little doubt on that score.[26]

Sherman, marveling at the general's supreme faith in the Navy's ability to achieve victory at Inchon, took him aside for a private conversation which the CNO later recorded in his diary:

> Had a long talk with MacArthur alone. He praised the Navy and spoke in glowing terms of its future. Said the entire Pacific should eventually be commanded by a naval officer. He agreed to my proposal that in the event of general hostilities, [Pacific Fleet Commander Admiral Arthur] Radford should move forward to wherever MacArthur is and take operational command of the entire Pacific Fleet. Criticized Air Force and blamed them for poor support of troops. Told me Marines were superb, but had a tendency to gripe. I told him again that Inchon was a dangerous enterprise if any resistance developed. He agreed it could be done only if there were none.[27]

Won over by MacArthur's genius, Sherman took his leave of the general who remarked to him in passing that the Navy was the "only service which hadn't let out a peep" of complaint.[28] As soon as he reached his office in Washington, Sherman immediately wrote to

MacArthur that "I feel much better equipped to cope with problems connected with the Far East because of my visit and the opportunity to talk with you."[29] And, of course, MacArthur was correct; the assault in mid-September took Inchon by surprise, broke the back of the North Korean Army, and sent it reeling in retreat back across the 38th parallel.

For the second and larger problem, keeping the war from expanding to the Chinese mainland and thereby risking full-blown retaliation by the Red Chinese and Soviet Russia, MacArthur proclaimed his maritime strategy in which the United States (or United Nations) would maintain its ring of bases that girdled the periphery of East Asia. In a letter to a Veterans of Foreign Wars convention in late August he called this ring the "island chain . . . from the Aleutians to the Marianas . . . from [which] we can dominate with air power every Asiatic port from Vladivostok to Singapore and prevent any hostile movement into the Pacific. . . . If we hold this line we may have peace—lose it and war is inevitable."

At the heart of the chain lay Formosa, "an unsinkable aircraft carrier and submarine tender" that possessed "a concentration of operational air and naval bases potentially greater than any similar concentration on the Asiatic mainland between the Yellow Sea and Strait of Malacca." He also looked upon the Nationalist Chinese as a potential manpower reservoir for fighting on the continent. The Truman administration did not agree with the importance of Formosa, nor the possibility of using its Nationalist Chinese troops, and the President had MacArthur withdraw his already published statement, while the JCS placed restrictions on MacArthur's movements into North Korea so he would not antagonize the Communist Chinese or Russians, restrictions that MacArthur modified to his own liking. The President therefore flew to Wake Island in October to meet MacArthur and to assert his authority over his Far East commander.[30]

At Wake MacArthur conveyed to Truman erroneous intelligence estimates that Red China would not enter the war in Korea and thus was faced with embarrassing reassessments when indeed the Communist Chinese crossed the Yalu the very next month, November, and drove back the United Nations ground forces. MacArthur now began not only to fear for the ability of his limited ground forces to hold in the South in the purely defensive war that the Truman administration wanted, but he also began to doubt whether his skeleton forces in Japan could resist a Russian attack on that country. He asked that National Guard divisions be called up to reinforce his Japan garrison, and then chafed when the JCS late in December told him to evacuate Korea if he could not hold against the Communists there with his available forces.

Douglas MacArthur had been through another such evacuation in the past—the Philippines nine years before. The nightmares of Bataan and Corregidor had led to the need for the long drive back across the Pacific, and MacArthur could never believe his situation at the end of 1950 paralleled the dark days of early 1942. Consequently he invoked his maritime strategy that had worked so well against Japan and responded to the JCS on 30 December.

He called for four measures to repel the Communist advance: (1) a naval blockade of mainland China, (2) air and naval attacks on the Chinese war-making industry, (3) the use of Nationalist Chinese troops in Korea inasmuch as the U.N. could not provide the necessary manpower, and, (4) a not altogether successful strategic ploy most recently demonstrated at Dieppe in 1942, namely the creation of a "diversion" by allowing the Nationalists to make amphibious raids on the mainland at least to promote guerrilla action there and even "possibly leading to counter-invasion."

These actions he believed would save the U.N. military position in Korea until a political decision was forthcoming "as to whether to maintain the fight in that area or to effect a strategic displacement of our forces with the view to strengthening our defense of the Littoral Island Chain while continuing our naval and air pressure upon China's military potential." To evacuate the Korean peninsula "without taking military measures against China proper," he believed, would hurt American prestige, encourage a Communist assault on Japan, and thus require heavy reinforcements for the defense of that country, particularly against Russia.[31]

The Joint Chiefs rejected MacArthur's recommendations as at least premature and at most alarmist. The Truman administration reflected the attitude of the majority of the American people in not wanting to risk expanding the war, which the JCS felt MacArthur would be doing with his desire and strategy to destroy Red China's warmaking capacity. Furthermore, they noted that the sudden reestablishment in December of a strong U.N. defensive position in the South was evidence that the military situation was being stabilized.[32]

Nevertheless, MacArthur the military commander clung to his maritime strategy. At no time did he advocate projecting the U.S. Army or Marines into Chinese territory; as he often said, "Anyone in favor of sending American ground forces to fight on Chinese soil should have his head examined." Rather, he would isolate the Korean peninsula from outside supply in classic maritime strategic fashion. Then, as he said in March 1951, he would go after Chinese supply lines: "Control of the sea and the air, which in turn means control over supplies, communications and transportation, are no less essential and decisive now than in the past."[33]

After regaining the Seoul line just south of the 38th parallel, as he recalled in his memoirs,

> I would then clear the enemy rear all across the top of North Korea by massive air attacks. If I were still not permitted to attack the massed enemy reinforcements across the Yalu, or to destroy its bridges, I would sever Korea from Manchuria by laying a field of radioactive wastes— the by-products of atomic manufacture—across all the major lines of enemy supply. . . . Then, reinforced by Nationalist Chinese troops, if I were permitted to use them, and with American reinforcement on the way, I would make simultaneous amphibious and airborne landings at the upper end of both coasts of North Korea, and close a gigantic trap. . . . It would be something like Inchon, but on a much larger scale.[34]

Unfortunately for MacArthur, his completely sound military strategy involved so many political aspects—not least the new idea of using anything nuclear—that the Truman administration could simply not entertain the possibility of adopting his plan. Equally tragic, the JCS had lost confidence in MacArthur's ability to restrain himself should the Communists launch a massive air attack from Manchuria; the Chiefs concluded he would probably retaliate with air strikes on China's Shandong (Shantung) peninsula. With none of his civilian or military superiors trusting his leadership, MacArthur faced removal from command—done finally in April when he made critical political remarks about the Administration's prosecution of the war in a letter to Congressman Joe Martin.[35]

Politics aside, Douglas MacArthur ended his career as America's foremost amphibious general. From his initial strategic ideas of 1942 through the Pacific victory and into Korea, he had matured into a maritime strategic thinker of the first rank. And so, in his final statement before a joint session of Congress in April 1951, he repeated his strategic views of the previous August, that the Pacific is

> a protective shield for all of the Americas. . . . We control it to the shores of Asia by a chain of islands extending in an arc from the Aleutians to the Marianas. . . . From this island chain we can dominate with sea and air power every Asiatic port from Vladivostok to Singapore and prevent any hostile movement into the Pacific. Any predatory attack from Asia must be an amphibious effort. No amphibious force can be successful without control of the sea lanes and the air over those lanes. . . . With naval and air supremacy and modest ground elements to defend bases, any major attack from continental Asia . . . would be doomed to failure. . . . Our line of defense is a natural one and can be maintained . . . [as] an invincible defense against aggression.[36]

General MacArthur's consistent advocacy of a maritime strategy for the United States in the Pacific places him alongside the nation's leading admirals who have similarly advocated such a strategy. Indeed, if one can excuse this bit of hyperbole (which may well cause him to turn over in his grave!), history would not be far wrong in remembering him also as *Admiral* Douglas MacArthur.

NOTES

1. Sir Julian Corbett, *Some Principles of Maritime Strategy* (New York: AMS Press reprint, 1972). The term "indirect approach" is usually attributed to Sir Basil Liddell Hart during the inter-World War period.
2. Clark G. Reynolds, "American Strategic History and Doctrines: A Reconsideration," *Military Affairs*, December 1975, p. 185. For King, see my "Admiral Ernest J. King and the Strategy for Victory in the Pacific," *Naval War College Review*, Winter 1976, pp. 57–64.

3. Quoted in D. Clayton James, *The Years of MacArthur* (Boston: Houghton Mifflin, 1970, 1975), v. I, 94:; Douglas MacArthur, *Reminiscences* (New York: McGraw-Hill, 1964), p. 32.
4. Maurice Matloff, *Strategic Planning for Coalition Warfare, 1943–1944* (Washington: Chief of Military History, 1959), pp. 186–191, 207–208.
5. His repeated attempts to gain control of Navy vessels are best described in Daniel E. Barbey, *MacArthur's Amphibious Navy* (Annapolis: U.S. Naval Institute, 1969), while in his forthcoming study of the War Shipping Administration Jeffrey J. Safford has discovered similar efforts with respect to the merchant marine. MacArthur "wreaked havoc with the WSA's turnaround and loading efficiency programs. At certain points he was confiscating the vast majority of the vessels the WSA sent to the South Pacific—for his own 'cross-trades' and for storage purposes." Safford to the writer, 7 May 1979.
6. Clark G. Reynolds, *The Fast Carriers: The Forging of an Air Navy*, rev. ed. (Huntington, N.Y.: Krieger, 1978), pp. 116–118; George C. Kenney, *General Kenney Reports* (New York: Duell, Sloan and Pearce, 1949), pp. 346–348; and Matloff, p. 455.
7. James, v. II, pp. 522–524; Reynolds, *Fast Carriers*, pp. 142–143, 243–245; Matloff, pp. 480–482; and Robert Ross Smith, "Luzon versus Formosa," in Kent Roberts Greenfield, ed., *Command Decisions* (Washington: U.S. Dept. of the Army, Office of Military History, 1960), pp. 465–469.
8. James, v. II, 539–541; Reynolds, *Fast Carriers*, pp. 245–246, 248–249; Matloff, pp. 486–487; and Smith, pp. 467–469.
9. Louis Morton, *Strategy and Command* (Washington: Office, Chief of Military History, 1962), pp. 668–669; Matloff, p. 487; and Reynolds, *Fast Carriers*, pp. 248–249, 322–323.
10. Forrest Sherman, "Notes on Conferences with CinCSWPA," 3-10 November 1944, p. 11, Sherman papers, Operational Archives, Naval Historical Center. On Japanese strength, see David James, *The Rise and Fall of the Japanese Empire* (New York: Macmillan, 1951), p. 246.
11. Sherman, p. 8. As for Chusan, "He seemed interested, but noncommittal." For the details, see Reynolds, *Fast Carriers*, pp. 323–324.
12. Walter Millis and E. S. Duffield, eds., *The Forrestal Diaries* (New York: Viking Press, 1951), p. 31. Entry of 28 February 1945.
13. George A. Lincoln, quoted in James, *MacArthur*, v. II, p. 764.
14. See my King essay.
15. Quoted in James, *MacArthur*, v. II, pp. 764–768, 770.
16. James, *MacArthur*, v. II, p. 359.
17. Sherman, p. 4.
18. Ibid., pp. 6-7.
19. Ibid., pp. 9, 12.
20. Reynolds, *Fast Carriers*, pp. 288–289, 298, 342–343, 368; Frank Futrell and James Taylor, "Reorganization for Victory," in W. F. Craven and J. L. Cate, eds., *The Army Air Forces in World War II* (Chicago: University of Chicago Press, 1953), v. V, p. 690; and Sherman, p. 8. For MacArthur's education about tactical air and naval gunfire in the Pacific, see James, *MacArthur*, v. II, pp. 281–282 and 283. For the details of OLYMPIC, see K. Jack Bauer and Alvin C. Coox, "OLYMPIC vs. KETTSU-GO," *Marine Corps Gazette*, August 1965, pp. 32–44.

21. James, *MacArthur*, v. II, pp. 773–774, noting MacArthur's press conference 3 days before the anticipated Russian attack. In his *Reminiscences*, p. 261, MacArthur simply chose to forget his advocacy of the Russian invasion because of his displeasure over the Yalta "sell-out." "From my viewpoint, any intervention by Russia during 1945 was not required." He allowed only that he had advocated it in 1941! See also William Manchester, *American Caesar: Douglas MacArthur, 1880–1964* (Boston: Little, Brown, 1978), pp. 438–439.
22. *Hearings on . . . The Relief of General . . . MacArthur* (Washington: U.S. Govt. Print. Off., 1951), Part I, pp. 57–58, quoted in Theodore Ropp, *War in the Modern World* (New York: Collier Brooks, 1962), pp. 381–382.
23. "JCS Visit to the Pacific, February 1950," prepared by the staff of the Commander in Chief Pacific Fleet/Admiral Arthur W. Radford/, citing JCS 1380/75 of 19 December 1949, JCS 1483/50, NCS 13/3 and NSC 49, Sherman papers. Also, *New York Times*, 14 February 1950.
24. Omar N. Bradley, "Notes on the Visit of the JCS to the Far East," February 1950, and MacArthur to JCS "Memorandum on Formosa," 14 June 1950, Bradley papers, National Archives, Box I.091. Charles M. Cooke to Forrest P. Sherman, 14 April 1950, Sherman papers. For the detailed sequence of JCS actions over Korea, see James F. Schnabel and Robert J. Watson, *The History of the Joint Chiefs of Staff: The JCS and National Policy: Vol. III, The Korean War* (Washington: JCS Historical Division, March 1979; also Wilmington, Del., 1979).
25. Sherman, "Memorandum for the Record" (actually an abbreviated diary) recording all JCS actions between 25 June 1950 and 11 April 1951, entries of 1, 8, 10, 15 and 21 August 1950, Sherman papers; and MacArthur, p. 341.
26. MacArthur, p. 349; Sherman, "Memorandum," entries of 21 and 22 August 1950; and M. W. Cagle and F. A. Manson, *The Sea War in Korea* (Annapolis, U.S. Naval Institute 1957), pp. 75–76.
27. Sherman, "Memorandum," entry of 23 August 1950; MacArthur, pp. 347–348, 350; Manson and Cagle, p. 76; David Rees, *Korea: The Limited War* (New York: St. Martin's Press, 1964), pp. 82–83.
28. Quoted in Sherman, "Memorandum," entry of 24 August 1950. See also Robert Debs Heinl, Jr., *Victory at High Tide: The Inchon-Seoul Campaign* (Philadelphia: Lippincott, 1968), pp. 38–41.
29. Sherman to MacArthur, 25 August 1950, Sherman papers.
30. Quoted in *New York Times*, 29 August 1950; MacArthur, p. 341; Rees, pp. 115ff.; Lawrence J. Korb, *The Joint Chiefs of Staff* (Bloomington: Indiana University Press, 1976), pp. 143–145. In the afterglow of Inchon, Bradley wrote to MacArthur on 29 September 1950 that "we want you to feel unhampered tactically and strategically to proceed north of the 38th parallel," which assurance he liberally interpreted. Bradley papers, Box I.091.
31. MacArthur to the JCS, 30 December 1950, Sherman papers; Sherman, "Memorandum," entries of 2, 9 and 10 January 1951; Rees, pp. 179–181; MacArthur, pp. 378–380. An interesting sidelight to his strategy, in view of the subsequent events in Vietnam, was MacArthur's view of the French crisis in Indochina. At the Wake meeting he told the President: "The French have 150,000 of the best troops there with an officer of the highest reputation in command. . . . I cannot understand why they do not clean it up—we have seen a debacle. . . . I cannot understand it." To which the Pacific Fleet commander Radford observed that the French had no

popular backing in Indochina and proposed that they raise native troops. Bradley noted that Truman and MacArthur agreed on the importance of Formosa. Bradley, "Verbatim Notes of Wake Island Meeting," p. 8, Bradley papers, Box II.

32. JCS to MacArthur, 55234 of 9 January 1951; Sherman, "Memorandum," entry of 19 January 1951, Sherman papers. For the dilemma MacArthur faced, see J. Lawton Collins, *War in Peacetime: The History and Lessons of Korea* (Boston: Houghton Mifflin, 1969), pp. 248–254.

33. MacArthur, pp. 387, 389.

34. Ibid., p. 384. Collins, pp. 215–216, believed MacArthur placed too much faith in air power.

35. Sherman, "Memorandum," entry of 8 April 1951; Omar Bradley statement to JCS of 23 April 1951. Copies in Bradley and Sherman papers. On the JCS reasons for MacArthur's relief, see Collins, pp. 284–285, and Manchester, p. 641.

36. Quoted in MacArthur, p. 401.

9

USES AND MISUSES OF WORLD WAR II MARITIME STRATEGY TODAY

Because of the adoption of "massive retaliation" with thermonuclear weapons as the basis for the American defense posture of the 1950s, followed by the debacle of American involvement in the Vietnam War during the 1960s and 1970s, the World War II maritime strategic thinking exemplified by Admiral King and General MacArthur fell into neglect for an entire generation. Only with the appearance of Secretary of the Navy John F. Lehman's "maritime strategy" during the early 1980s did the United States begin to reexamine historical American strategic activities.

This chapter is a reminder to historians and strategists alike of not only what American, British, Chinese, and Japanese leaders preached and practiced as maritime strategy but to suggest how World War II maritime strategy might be applied or not applied to the shaping of current American defense and foreign policy. Originally presented as a paper before the Department of Strategy faculty at the U.S. Naval War College in 1985, it was published as "The Maritime Strategy of World War II: Some Implications?" in the May–June 1986 issue of the *Naval War College Review*.

History and the Sea

In its reorganization on 1 February 1941, the U.S. Fleet was proclaimed to be a two-ocean navy. This marked the first step in an ambitious program to make the two-ocean navy a reality—one whose Atlantic and Pacific fleets would be powerful enough to defeat both Germany and Japan at sea. Today, the U.S. Navy is a five-ocean force with global responsibilities but with only one major superpower enemy, the Soviet Union.

The years 1943–45 were epochal in the strategic life of the United States in general and in particular for the Navy because of the dual victories in two oceans. The advent of nuclear weapons at the end of the global conflict tended to obscure any applicable "lessons" for the future, at least until the post-Cuban missile crisis era. As the unsuitability of nuclear weapons have become more apparent as a means for warfighting since 1962, the importance of World War II experiences has taken on new meaning. Thus, it behooves contemporary strategists to take a hard look at the strategic realities and subsequent historical perceptions of the wartime era in order to understand which of them are useful and which are not in the shaping of our own strategic policies.

Like the strategy makers of World War II, historians have tended to view the conflict in the plural. That is, the European and Pacific theaters have been regarded as two separate conflicts. And for all intents and purposes they were, save for the tight logistical and landing craft requirements that had to be met on a global scale. Yet in fact, Anglo-American strategic policy may be viewed in the singular, even if both powers in that period and in the prewar years insisted on nationalistic unilateral goals and activities. Specifically, the British Commonwealth and the United States each acted from national motives, but both practiced maritime strategies which came together during World War II.

A maritime strategy is one ascribed to a sea-oriented great power—a nation geographically isolated, democratic, with a mercantile economy and a navy as the "senior" armed service, the first line of defense. In a major conventional war with another great power, such a maritime state seeks:

Reprinted by permission, *Naval War College Review*.

- to protect its own ocean-borne commerce by maintaining command of the vital SLOC (sea lines of communications);
- to isolate the enemy homeland; and
- *most importantly,* to support and augment the army of its major ally on the continent. The latter is the primary force that must engage and defeat the enemy army on the continental mainland in order to win the war.

This is a strategy of concentration through neutralizing or defeating the enemy at sea and in the air while hitting at his strategic center—his home armies—with a powerful allied army.

Britain generally practiced this maritime strategy in her wars with France and Germany between the Seven Years War of the 1750s and World War I. Her notable failures were the War of American Independence, when she could never obtain a continental ally to oppose the armies of France, and World War I, when the Imperial Russian Army failed on the Eastern front then succumbed to internal revolution. During the 1920s and 1930s Britain looked to France to hold the line in Europe, only to watch Hitler defeat her ally in 1940. Not until Hitler invaded Russia in June 1941 could the British hope to apply their traditional maritime strategy in Europe.

The United States, not becoming a major military power until World War I, generally left European leadership and strategy to the British and instead concentrated on the Pacific and the Japanese menace. War Plan Orange, initiated in 1907 and refined during the 1920s and 1930s, was by no means a maritime strategy. Rather, it was purely a *naval* strategy. That is, both the Navy and Army joint planners intended to move across the Pacific to defend, or retake if necessary, the Philippines from an aggressive Japan. This meant defeating the Japanese fleet in battle and utilizing the Army for defense of the islands and the new Fleet Marine Force as the amphibious component.[1]

The ultimate goal of Plan Orange was the sea-air blockade of Japan's home islands until they surrendered. Unfortunately, this naval strategy did not take into account the facts that Japan was embarking on a strategy of continental conquest of the mainland of Asia, under the guise of "the Greater East Asia Co-Prosperity Sphere"; that Japan's perceived ultimate enemy was Soviet Russia; that Japanese

History and the Sea

strategy was dominated by the Army; and that the Imperial Japanese Navy was a defensive fleet oriented to supporting the Army on the continent. With the seizure of Korea in 1910 and Manchuria in 1931, the invasion of China proper in 1937 and subsequent move into Indochina, Japan should have been perceived as the *continental* power she in fact was. Awed by Japan's impressive navy, both the U.S. Army and Navy misperceived Japan's primary intentions and strategy—the conquest of China and the defeat of Russia, and not an eastward thrust toward Hawaii and the vast Pacific.[2]

The dramatic events of 1941—America's undeclared involvement in the Battle of the Atlantic, the German invasion of Russia, and finally Pearl Harbor—brought the British Commonwealth and the United States together in common cause and common strategy. Despite differences of emphasis and details, the Anglo-American Combined Chiefs of Staff embraced a Maritime Strategy. True, the British predominated in strategic planning for the European Theater of Operations (ETO), while the Americans got a free hand in the Pacific, but in essence they both advocated a *singular strategic policy* in both theaters. This called for three things:

• The protection of their ocean-borne commerce by winning control of the vital SLOC. This meant defeat of German surface raiders and the U-boat on the one hand, and the destruction of the Japanese Combined Fleet on the other.

• The isolation of the enemy homelands. In Europe, this entailed seizing control of the Mediterranean by regaining North Africa and taking Italy out of the war to clamp a blockade on the continent, and winning control of the air over the continent. In the Pacific, this required the capture of key islands for the development of advanced naval and air bases, the submarine offensive against the Japanese merchant marine, an aerial mining campaign to close the harbors of Japan, and winning control of the air over the entire Western Pacific and Japan proper. In sum, this was the sea-air blockade of the Japanese Empire as loosely envisioned by War Plan Orange.

• *Most importantly*, the supply, support, and augmentation of Allied armies on the continents of Europe and Asia to enable them to defeat the huge armies of Hitler and Tojo upon which the dictatorships were based. Once these Allied armies on the two continents took the strategic offensive, then the maritime powers could project their own though relatively much smaller armies onto the continents, in the enemy's rear, to help hasten the collapse of the fascistic armies and their political systems.

In the ETO, there was no doubt that the main fighting force on the ground was the Red army. Beginning with Hitler's invasion of Russia in the summer of 1941, Anglo-American navies and merchantmen braved the extreme hazards of the Murmansk run (as well as the safer passages to Iran and across the North Pacific) to sustain Soviet ground forces until they were able to turn the tide after Stalingrad in 1943. As the Red counterattack threw the Germans on the strategic defensive, the Atlantic Allies could begin planning for their landing in the German rear—the second or Western front long desired by Stalin. When it came, at Normandy in June 1944, the Wehrmacht was in full retreat along the Eastern front. By the autumn 300 Russian divisions (5,500,000 men) were driving back 150 German divisions (4,100,000 men) in the greatest land war in history. By contrast, the Anglo-American effort in the West consisted of 38 divisions (1,000,000 men) against 41 understrength German divisions (though more German divisions were tied down elsewhere: 24 in Italy, 17 in Scandinavia, and 10 in Yugoslavia). True, the Nazi war machine was caught in a grand pincers, but the weight of Allied arms came from the East, took Berlin, and forced the German Army to its knees.

In the Pacific, little initial thought was given to the major striking force. In 1941–42, as in Europe, Allied strategy was essentially defensive: to hang on and to arrest the Japanese offensive, accomplished at the Coral Sea, Midway, and Guadalcanal. And once the counterattack was contemplated, both Admirals King and Nimitz merely resurrected the Orange Plan for driving across the Central Pacific toward the Luzon bottleneck. This involved taking the necessary islands for bases and instituting the long-planned sea-air blockade of the Japanese home islands. General MacArthur had no quarrel with this, except that he insisted the Philippines be liberated in the process. Chief of the Army Air Forces General H. H. Arnold bought it, since the capture of the Marianas meant airfields for his very long range B-29 bombers hitting Japanese cities.

Yet, in the Maritime Strategy, the American thrust into the Western Pacific was still the *holding* element, the isolating of the enemy homeland while the major Allied army defeated the powerful Japanese army on the continent of Asia. From the outset, this force was the Nationalist Chinese army controlled by Chiang Kai-shek, to whom American aid and volunteers had been directed since before Pearl Harbor. The master of American strategy in the Pacific, Admiral King, never wavered in his view of "Chinese manpower" being "the ultimate land force in defeating the Japanese on the continent of Asia. . . ."[3] While the millions of underequipped and poorly led Nationalist Chinese battled the two million superb troops of Japan, King planned for the traditional amphibious landing in the enemy's "rear" to trap Japan's armies on the continent.

Since the U.S. Central Pacific offensive could not reach the Chinese coast until the autumn of 1944, King tried to convince the British to launch a major drive through Burma in 1943–44. But the lack of resources led Britain to keep her efforts in the CBI theater on a smaller scale. Thus King could only pin his hopes on his own navy and assault forces reaching the coast of China for an Eastern-style Normandy. In his periodic planning conference with Nimitz in January 1944 he "stressed the point that all operations were aimed at a drive through the Pacific to China in order to exploit Chinese geographical position and manpower." He repeated this point in their next meeting two months later.[4]

The problems in King's maritime strategy was the Nationalist Chinese army. By early 1944 it had yet to turn the tide of battle against the Japanese army, a necessary prerequisite to any Allied landing on the coast of Asia. And as the year wore on, any confidence that the American planners had in Chiang Kai-shek gradually evaporated. If the Nationalist Chinese army could not provide the means to achieve victory on the mainland of Asia, what force could? One thing was certain, the U.S. Army could not. General MacArthur declared late in 1944 that the manpower requirements were simply too great for an American land war in Asia.[5]

The logical solution to fulfilling the Maritime Strategy in the Pacific was to bring in the Soviet army. In March 1944, even while calling for the use of Chinese manpower, Admiral King broached the idea to Nimitz. Since no transportation facilities then existed for the Americans supplying the Russian Far Eastern provinces, King ordered Nimitz to study North Pacific SLOCs to the Asian mainland. Both men and their advisers repeated their interest in this possibility in May and again in their July meeting. They agreed that the superb Japanese (Kwantung) anmy in Manchuria had to be kept occupied. Nimitz preferred landing at Amoy on the Chinese mainland opposite Formosa in order to supply the Chinese, "thus utilizing Chinese manpower as the ultimate land force in defeating Jap Forces on the continent of Asia." He then wanted an advance up the coast to Shanghai, the only decent deep-water port for handling such vast supplies. King was less certain and thought that the Kuriles and North Pacific SLOC might be better, for should Russia enter the Pacific war "we shall be that much to the good."[6]

However, the failure of even the best 35 Nationalist Chinese divisions raised serious doubts about their ability to help defend such large U.S. coastal enclaves, which were in effect "islands," said Nimitz—on the Chinese seaboard—which led King to decide on landing on the real island of Formosa instead of coastal Amoy-Shanghai. Unfortunately, by the time MacArthur was ashore at Leyte in the southern Philippines, and King and Nimitz met again in November 1944, the Nationalist Chinese were retreating before a Japanese offensive. And U.S. forces were simply not available for an assault on either Amoy or Formosa. Instead MacArthur would go next to Luzon in the northern Philippines, and plans were initiated for an ultimate American invasion of Japan proper one year hence.

But none of the Pacific leaders wanted to undertake such an invasion unless the Japanese army in Manchuria was pinned down first. Thus King was able to inform Nimitz in November 1944 that Stalin would enter the Pacific war after he had had three months to stockpile arms and manpower for no fewer than 60 divisions— meaning, three months after Germany quit.[7] MacArthur told Secretary of the Navy Forrestal in February 1945 "that the United States could put 13 divisions into Kyushu then 22 into Honshu as 'the *coup de main*' from the rear [but only] while substantial portions of the military power of Japan were . . . heavily engaged by the Russians in Manchuria."[8] In fact, however, neither he nor the admirals really wanted to undertake the invasion of Japan and believed, or at least hoped, that Japan would quit before it became necessary.

Therefore, the key to victory remained the Red army defeating the main Japanese field army, the political edifice upon which Japan's wartime government rested. Other forces were also at work simultaneously to undermine that imposing army. In a masterful stroke of strategic deception, U.S. intelligence teams so effectively kept active bogus radio traffic calling for a Formosa-Amoy landing that substantial numbers of Japanese troops were shifted from Manchuria to the threatened coastal regions by the spring of 1945.[9] Even more important, where Chiang Kai-shek had failed in the south of China, Mao Tse-tung's Communist Chinese army thoroughly demoralized and weakened the Japanese army in north China using guerrilla tactics over the years. Finally, the defeat of major Japanese army forces on Luzon and Okinawa during the first half of the year forced Japan's generals to finally contemplate surrender.

The Maritime Strategy worked. True to his word, Stalin unleashed his 60-division army on the Japanese Manchurian army—three months after the German surrender, on 8 August 1945, two days after Hiroshima and the day before Nagasaki. The Red army swept over the once-vaunted Japanese ground forces in Manchuria and North Korea, giving the samurai generals good cause to quit once the Emperor ordered it. Of equal effect, on the 2,000,000 defenders of the home islands, were the two A-bombs, the final episode in the deadly air-sea blockade which had destroyed Japan's wartime economy.

Anglo-American strategy in World War II was a maritime one, aimed first and foremost at the defeat of the massive continental armies of Nazi Germany and Imperial Japan. In both instances, the Allied vehicles for victory in Europe and Asia were the sprawling armies of Soviet Russia and Communist China—strange bedfellows, to be sure, of the Western democracies but strategic necessities nevertheless. Their ultimate triumph on land was made possible by the other two key elements of the wartime Maritime Strategy: First, control of the SLOC in both oceans, and second, the sea-air isolation and bombardment of the enemies' homelands. The final seaborne invasion had proved important in Europe but unnecessary, as it turned out, in the Pacific.

World War II Maritime Strategy Today

All the foregoing occurred more than forty years ago, and inasmuch as history never repeats itself exactly, current strategic analysts must be highly selective in seeking to glean useful lessons or implications of past events such as this scenario of World War II.

1. Today there is one major superpower adversary rather than two. Like Nazi Germany and Imperial Japan, the Soviet Union is a totalitarian dictatorship whose main armed force is its army. The Red army not only defends the continental frontiers of the Soviet state but is an internal police force used to maintain order within Russia and its buffer satellite subject states. The destruction or implied destruction (deterrence) of the Red army must therefore lie at the root of contemporary Western strategy, including the Maritime Strategy.

2. The Russian navy traditionally has been a defensive adjunct to the ground forces. Like Germany's, this navy's primary role in conventional war has been oriented toward *guerre de course*, the interruption of Western SLOCs. But the Soviets have now developed a surface naval capability markedly similar to that of Japan's. If history—specifically, the continental dictatorships of World War II—is any guide, it is highly probable that the Red army will continue to dictate naval doctrine and strategy. Therefore, it would be prudent to take this into account when shaping our current Maritime Strategy.

3. Inasmuch as major Allied continental armies acted as the key agents for victory in the Anglo-American Maritime Strategy of 1943–45, a similar alliance or strategic association would seem to be an essential prerequisite for our own Maritime Strategy. If we treat the Soviet menace as we did that of the Axis powers, as a two-theater affair, then we must cultivate separate allied armies to provide our continental hitting power on both ends of the Eurasian landmass. In Asia, the comparison with World War II is useful; the huge Chinese Communist army is a potential ready-made solution. Europe is a different matter, for U.S. ground forces comprise only a part of the many-nation NATO coalition, whose combined ground forces are still greatly outnumbered. Hence, a strict comparison with the Maritime Strategy of World War II is less useful. Or, it may be that our Maritime Strategy should be global—using the bulk of the Chinese army to attack on such a large scale in Asia as to affect Russian dispositions in Europe. In any case, a key lesson of the continental

History and the Sea

Allies of World War II is, however much their successors today are an absolute essential ingredient of the contemporary Maritime Strategy, their cooperation and success cannot be guaranteed—witness the debacle of Chiang Kai-shek's Nationalist China. Point: never trust or depend on an ally.

4. Finally, the Maritime Strategy of 1943–45 took a long time to reach fruition. Maritime strategies in all eras have been long, slow and subtle processes, a protracted struggle which wears away at the enemy dictatorships. In this sense, time is on the side of the maritime powers. For Soviet Russia—like Imperial Russia with the Bolsheviks, Nazi Germany with the Resistance, and Japan's Manchurian army with Mao's guerrillas—must maintain internal police control over whatever restive patriots it conquers. Therefore, in spite of possible Dunkirks and even temporarily interrupted SLOCs à la the Murmansk run, the United States must impress upon the Soviet regime that we will persevere even in the face of possible initial reverses. This can be done by following a Maritime Strategy with which to undermine and destroy the Russian Communist political-military system as we did those of the fascists.

Still, each historical event is unique, and, generally, contemporary strategy cannot be based on historical precedents. The Maritime Strategy of today can be the exception, provided that the historical experience of World War II is studied and tested continually and in detail for clues into its implications for our time. For in strategy making, the greater danger than a complete ignorance of history is its *mis*application.

NOTES

1. Edward S. Miller, "War Plan Orange, 1897–1941: The Blue Thrust Through the Central Pacific," Unpublished paper, U.S. Naval Academy Naval History Symposium, 26 September 1985.
2. John Toland, *The Rising Sun: The Decline and Fall of the Japanese Empire, 1936–1945* (New York: Random House, 1970), pp. 7ff., 42ff., 66, 74–80; my "The Continental Strategy of Imperial Japan," U.S. Naval Institute *Proceedings*, August 1983, pp. 65–71.
3. Ernest J. King and Walter Muir Whitehill, *Fleet Admiral King* (New York: Norton, 1952), pp. 419–420, 362, 541–542; my "Admiral Ernest J. King and the

Strategy for Victory in the Pacific," *Naval War College Review*, Winter 1976, pp. 57–64.

4. King-Nimitz Conference minutes of 3 January and 6 March 1944. E. J. King papers, Series IV, Box 10, Naval Historical Center.

5. MacArthur quoted by Rear Admiral Forrest Sherman, November 1944, p. 11, in minutes of their meeting. Forrest Sherman papers, Naval Historical Center. Also my "MacArthur as Maritime Strategist," *Naval War College Review*, March-April 1980, pp. 79–91.

6. King-Nimitz Conference minutes, March, May, July, September 1944.

7. King-Nimitz Conference minutes, November 1944.

8. MacArthur quoted in Walter Millis and E. S. Duffield, eds., *The Forrestal Diaries* (New York: Viking, 1951), p. 31, entry of 28 February 1945.

9. Katherine Herbig, "The Impact of Strategic Deception in the Pacific: Operation Bluebird," Unpublished paper, U.S. Naval Academy Naval History Symposium, 26 September 1985.

10

EIGHT CENTURIES OF CONTINENTAL STRATEGY: IMPERIAL AND SOVIET RUSSIA

Before the United States and its allies dare apply a Maritime Strategy against the Soviet Union, as suggested in the previous chapter, Russia must be examined and appreciated strategically as the continental power she has been over the course of eight centuries. Like Imperial Japan, discussed in chapter 6, Russia has taken to the oceans as a naval power only in the short recent period of her history—since the 1960s. Not until analysts use the long view of Russian history, Imperial as well as Soviet, can they begin to interpret Russian strategic behavior realistically.

The Soviet Union has inherited from early twentieth century Germany and Japan the mantle of America's prime enemy. Before the American entry into World War I, the U.S. Navy's strategic planners believed that should the British navy be defeated, the Atlantic Fleet would have to repel the advancing German High Seas Fleet from our Eastern seaboard. And in the early days of World War II, with the crushing defeat of the Pacific Fleet at Pearl Harbor, a descent by the Japanese Combined Fleet on the West coast seemed a likely possibility. As far-fetched as these possibilities appear through hindsight, they were regarded as very real threats at the time, as the literature and documents of the day attest.

So now we have the Russians. Although the meaning of "invasion" is vastly different in the nuclear age than in the past, the fears of the

Eight Centuries of Continental Strategy

American people and their strategic planners over a potential sea and air attack are no less real than were those of our forebears as they regarded Germany and Japan. The literature and highest level policy planning leave no doubt of that. Yet, aside from thermonuclear weapons, just how great is the threat of today's Russian navy and army to America's well-being and security? The strategy makers, for their part, must assume the worst; as a Canadian naval officer once remarked to me, "We are paid to be pessimists." Preparedness is the job of the military professional. But it must be based on strategic realities, not shadowy fears.

If the strategic historian can make any contribution to analyzing the threat of the Soviet Union, it is in his ability to use the wide-angle lens of his profession to compare and contrast late twentieth-century Russia with the great continental powers of the modern period, the past four hundred years, and to understand Russian strategic doctrines stretching across nearly an entire millenium. With a bit of imagination, the strategic historian ought to place himself fifty or one hundred years in the future in order to look back and discern the factors of continuity and change in the Russian strategic experience. This unique perspective can then be added to the general contemporary discussion over the strategic nature of Russia in our own time. Such is the intent of this chapter, written especially for this volume and before the advent in 1985 of Mikhail Gorbachev as General Secretary of the Soviet Communist Party and his policies of *perestroika* (restructuring) and *glasnost* (openness).

The British bluejacket, observed naval writer Fred T. Jane in 1899, "believes that every Russian sailor lives chiefly on tallow-candles varied with bear's grease, decayed fish, and soap. He believes that Ivan fears but one thing—cold water. He further believes that Ivan is a person of no spirit at all in the ordinary way—that it has 'all been flogged out of him.' . . . Ivan is a big, strong, burly fellow with a sluggish good temper—like a big Newfoundland dog.

"He is simple and childish, and his intelligence is not high. He is amenable and willing, anxious to do his best and to find fun in his profession in his own melancholy way. . . . Ivan realizes that he exists *to be shot at*; Jack [the British tar, on the other hand, believes] that he exists *to shoot at others*, and this psychological difference is as heavy a one as can well be:—it is all the difference in the world."[1]

That the future publisher of *Jane's Fighting Ships* was not far wrong was dramatically demonstrated six years after he wrote this by the virtual annihilation of the Russian fleet by the Japanese at the Battle of Tsushima Straits. The question is, was the disaster at Tsushima typical of Russia at sea—or an exception? Put another way, was Tsushima a fair indication of Russia's national and military prowess and skill, then or now?

Alfred Thayer Mahan thought it was typical. "Militarily, Russia as a nation is not enterprising," the great savant of sea power wrote in 1911. "She has an apathetic bias towards the defensive" and therefore utilized her navy in the imperial tsarist period as a "Fortress Fleet . . . a dominant conception in Russian military and naval thought." In fact, he considered the notion of the Fortress Fleet to be "distinctly Russian," reflecting "national temperament; that is, national characteristics, national bias. For, for what does Fortress Fleet stand? For the defensive idea. . . . In what kind of warfare has Russia most conspicuously distinguished herself? In defensive." He cited the Russian repulse of Napoleon in 1812 and the stand against Britain and France in the Crimean War of 1854–56 as examples, to which we can add the defense against Hitler in 1941–42. "[By] virtue of her territorial bulk and vast population, she has, so to say, let the enemy hammer at her, sure of survival [by] virtue of mass."

Such an attitude was not necessarily the fault of the admirals, said Mahan. "The Russian Naval General Staff clamored for command of the sea; but in influence upon the government, the responsible director and formulator of national policy, it did not possess due weight." The tsarist government laid "all stress on the fortress, making the fleet so far subsidiary as to have no reason for existence save to help the fortress." Russia threw "national defense for the coast lines upon fortifications only. . . ." Thus the Russian Pacific Squadron remained anchored under the guns of Port Arthur, China, until it was lured out and defeated in 1904, and the Baltic Fleet, whose normal mission was defense of the Baltic coast, steamed 18,000 miles to its doom at Tsushima in a vain attempt to save Port Arthur.[2]

Mahan's view has been shared by the Communist successors of the Imperial navy. "All attempts by Western propaganda to portray the Soviet naval presence in the World Ocean as an expansionist drive are absolutely groundless," proclaims a 1981 work by the Soviet navy. "The Soviet Armed Forces have never been used for aggressive purposes. All the wars the USSR has waged"—that is, since 1917— "have always been directed solely against those who encroached upon its integrity and the freedom and socialist gains of the Soviet people."[3]

To properly judge whether or not the Russian nation has been a strategically continental state, four qualitative historical factors need to be addressed: geopolitical, economic, cultural, and military. And the best means for examining these factors is to compare Russia with the other great continental powers of modern history which have also maintained formidable navies: France, Germany, and Japan. In all four cases, including Russia's, the characteristics of the strategically continental state turn out to be diametrically opposite to the historical traits of the great modern strategically maritime powers or thalassocracies. The Netherlands, Great Britain, and the United States have been strategic islands, politically democratic, economically capitalistic, culturally free and creative, and militarily dependent on standing navies and allied armies.

GEOPOLITICAL

Given the sprawling landmass of the continental state, a strong authoritarian government is the primary requirement for manning and maintaining a large standing army and fixed fortifications for *external* security. This means that the continental army is also a major instrument of civil authority, subjecting the will of the people to that of the state—a para-police force, repressing and suppressing the masses for purposes of *internal* security. Such subservient peasant-workers are impressed into the army like nothing so much as cannon fodder. By its sheer numbers of uniformed manpower, the massive army thus becomes a powerful political tool of the state, its marshals and generals major policy makers in shaping the political character of the government. Not only that, but the generals determine and protect the head of state, who is an authoritarian dictator. The navy and its admirals, considered to be an unnatural political entity, are but a minor political adjunct to the senior service, viewed with suspicion for their lack of political commitment and preference for moderation in international affairs.

Throughout history, all great continental powers have been characterized by just such a despotic political system—conservative, reactionary, totalitarian, militaristic. The royal monarchs held absolute control over their countries, kept in authority by an aristocracy of landowners who provided the statesmen and generals.

Seventeenth- and eighteenth-century France was the archetype absolute monarchy, its political base, typically, being control over the land with its royal Bourbon princes and barons. In spite of the "enlightened" liberal ideas of its *philosophes* who inspired the Revolution in the 1780s, France could not throw off its geopolitical straitjacket: to support the Revolution, the Directory turned to the military, namely Napoleon Bonaparte, who lost no time in proclaiming himself First Consul, then Emperor—in short, one of the greatest authoritarian rulers in history. The subsequent political history of modern France was one of seesawing between "restored" monarchs and several republics, culminating in the sordid Vichy collaboration with the Nazis that left France a minor power that even strongman Charles de Gaulle could not restore to great power status after World War II.

Modern Germany—Prussia before 1870—fit the continental geopolitical mold with unbroken authoritarian and militaristic zeal. The Hohenzollern monarchs, exemplified by Frederick the Great in the mid-eighteenth century, were fixed to the land and its many princely fiefdoms. To defend the inseparable sacred national elements of *Blut und Boden* (blood and soil) the general-kings had to adroitly maneuver their armies between their several exposed borders to insure the survival of the state. Ruled by the *Kaiser* (king), the later German *Reich* (empire) tried to dominate Europe by pushing eastward: the *Drang nach Osten*. Her aristocratic landed *Junkers* generals, notably Erich Ludendorff, took control of the state during World War I and even provided stability under the short-lived postwar Weimar Republic. Their legacy was seized by the new *Fuehrer* (leader), Adolf Hitler, who also sought *Lebensraum* (living space) by moving eastward against Russia. In defeat, Germany lost her great power status.

"The prototype of modern Germany," wrote Mahan in 1910, lay in the example of the Roman Empire of antiquity, inherited through the Holy Roman Empire of the later Middle Ages and early modern period. Like these predecessors, the new German state believed "in the subordination of the individual to the state" which "sinks the unit in the whole." In fact, he noted also "that the Japanese Empire . . . is manifesting the same restless need for self-assertion and expansion, comes to its present with the same inheritance from its past, of the submergence of the individual in the mass. It was equally characteristic of Sparta among the city states of ancient Greece. . . ."⁴

Like the warrior state of Sparta among the Greeks, Japan entered the modern world of the West dominated by a landowning aristocracy, the Tokugawa Shogunate, which had ruled Japan through military strength for two and a half centuries until the restoration of the Meiji emperor in 1867. The elite *samurai* warriors before and after the industrialization begun under the Meiji regime worshiped their ancestors by practicing the code of *Bushido*—glorious death in battle in the service of their Emperor. At the end of the nineteenth and beginning of the twentieth century, Japan sought to dominate the Asian continent by subduing China and expelling the Russians as well as other Western powers with colonial holdings, including the United States. Typical of continental powers, when threatened with military defeat or stalemate the Japanese turned to the military, namely General Hideki Tojo and his fanatical cohorts. The result was defeat in World War II and loss of great power status.

In all three cases, the role of the navy tended to lie outside the geopolitical mainstream of the continental-military state system. Like the ill-fated Duke of Medina Sidonia, the admiral who lost the Spanish Armada in 1588, the navies of Bourbon France, Hohenzollern Prussia/Germany, and Meiji Japan drew their leaders from the landed gentry. Many became proficient in the naval profession—the names of Suffren, Scheer, and Togo come to mind—but were subordinated to the will of the army. Jean Baptiste Colbert created a great French navy in the late seventeenth century, but his reforms were not sustained. The Royalist admirals resisted the French Revolution by turning over their Toulon fleet to the British in 1793, while poor Admiral Villeneuve committed suicide rather than face Napoleon after his crushing defeat at Trafalgar. Admiral Alfred von Tirpitz in 1900 developed his "risk theory" of heavy German naval construction to discourage Britain from risking war and disdained army ambitions on the continent. Finally, in 1917–18, the Fleet mutinied, while Hitler's admirals were less than enthusiastic Nazis. In Japan, the best naval leaders were those who like Isoroku Yamamoto counseled the Emperor to avoid war with America and then as moderates worked for its termination during that losing struggle.

The Russian geopolitical experience has differed from those of France, Germany, and Japan only in magnitude. It was and is an incomparably *greater and deeper* reality, rooted in the unique fact that Russia has not only been invaded repeatedly over the past one thousand years but, worse, that it was early on conquered and occupied for two and a half centuries by the Mongol Horde of Central Asia. A politically fragmented, constantly warring Medieval people, the Russians of Eastern Europe were unified under the common disaster of conquest by the brutal armies of Batu Khan between 1236 and 1240. These fierce equestrian Orientals decapitated every combatant who opposed them, burned the cities and villages in a campaign of sheer terror, then ruled the Russians through the native princes of the several cities. The Mongols meanwhile converted to Islam. The inexorable offensive sweep of the Mongols left an indelible impression on the conquered of the need for the defensive, which has governed Russian strategy ever since. The fact that the invader came from the East also helps to explain Russian respect for the armies of Imperial Japan and Red China in the present century.

Unlike the Americans, whose national experience is still little more than two centuries old, Russia has very ancient origins, stemming largely from the nationalistic spirit which finally enabled the Russians to expel the Mongols in 1480 after an occupation of 240 years! But though Tsar Ivan III ("The Great") supplanted Mongol rule, he imposed his own authoritarian controls that allowed the peasantry no greater freedom than had the Horde. The long occupation had the concurrent effect of isolating Russia from the progress of Western Europe. In the words of historian Robert W. Daly, "Russia was amputated from Europe of the Middle Ages and Renaissance, and the humanizing bases of justice and democracy common to the West are therefore missing in the development of Russia, which, to the contrary, acquired a tradition of absolute rule." In contrast to the individualism spawned by the maritime West, "Russia was oriented towards the East, not only in superficial customs and dress, but in fatalistic disdain for the value of a human life. Russians were taught to be submissive by fear. It is noteworthy that no Western people have voluntarily chosen the neo-Mongolism of communist rule."[5]

The rule of the tsar thereafter emanated from the Kremlin in Moscow, enforced by the landowning nobility of *dvoriani* who provided the officers for the main branch of Ivan the Great's standing army. The power of the tsar was further solidified under the reign of the ruthless Ivan IV ("the Dread" or "the Terrible"), 1530–1584. He created the merciless *Oprichnina*, 20,000 secret police who persecuted and expelled the *boyar* class of lesser nobles to consolidate Ivan's power but who proved so dangerous that he finally abolished the organization. But he maintained as an urban police the murderous *Strel'tsi*, the first military unit equipped with firearms instead of bows-and-arrows and which became a closed military guild. Thus protected from internal rivals, Ivan used his army to repel invasion attempts by the Crimean Tartars, descendants of the Mongols inhabiting the Black Sea region and which managed even to capture and destroy Moscow in 1571.

By the beginning of the Romanov dynasty in 1613 (which lasted 304 years, until the Revolution in 1917), the tsars wielded their absolute power to thwart invasions not only by the Tartars and Turks from the south but by mighty Poland and Sweden and eventually Prussia, France, and Germany from the west. All the while they had to put down revolts and attempted coups by subject border peoples and rivals to the throne. Russian society remained basically feudalistic and repressive even after Peter the Great, 1695–1725, attempted to bring his country into the modern world of the West. When the *Strel'tsi* revolted in 1698, Peter crushed them and substituted his own Royal Guards to protect the crown. So powerful did the Guards become that they alone decided the succession to the throne for nearly a century. A violent, supreme autocrat, Peter tailored his government and economy into a veritable war machine to support an army of more than one million men and more than one thousand naval vessels which he flung at enemies on his borders. His greatest success was defeating the invasion by King Charles XII of Sweden in 1709. Peter capped his considerable achievements by proclaiming himself Emperor in 1721. The next year he instituted the Table of Rounds, a merit system based on an individual's military, diplomatic, or judicial service to *the state*.

The lowly serfs, wedded to the soil of imperial "Mother Russia," blindly and stoically served their tsar or tsarina. Catherine the Great, tsarina from 1762 to 1796, sent them against the traditional Swedish and Turkish foes and, like all continental monarchs at the time, feared the liberal ideals of the aggressive French Revolution. To strengthen her frontier against this 'cancer,' she used her army to help Austria and Prussia partition Poland in 1795, thereby destroying it as a nation until 1919. Subsequent Russian wars against the French, which even spread into the Mediterranean, culminated in Napoleon's abortive invasion of Russia in 1812 and the ensuing retrenchment of all the monarchies in Europe.

The virtual military dictatorship of Nicholas I, 1825–1855, was so brutal that one million men in the army died from illness and neglect over two decades. The concurrent unrest in the army was complemented by no fewer than 700 peasant uprisings against the regime. During 1827–1832 several mutinies occurred within the navy, including all 6,000 sailors based at Sevastopol, who had to be crushed and either executed, imprisoned, or sent to Siberia with their families. Ever since Peter's reign, both the army and navy had relied heavily on foreign officers; by 1855 three-fourths of the top commanders in the army were Germans. Their performance in the Crimean War at that time was notably poor, but the backwardness of the Russian state and military required such Western-schooled leaders to enable Russia to fight Western enemies.

The so-called "emancipation" of the serfs in 1861 did little to introduce genuine liberalism into the archaic Russian imperial political system, even as the pan-Slavic movement led the Kremlin to seek footholds in the Balkans and to more fighting with the Turks. The suppression of serfs and soldiers continued under Tsar Alexander III who in 1882 closed the theretofore open military court; now held in secret, it could punish political offenders in the manner of the typical police state. Imperial Russia still had a long way to go to catch up with the West, then experiencing the rise of the common man on the Anglo-American model.

Thus, a German officer could observe in 1889, in a passage resembling the one of Fred Jane ten years later quoted at the start of this essay:

> . . . The Russian is matchless when the necessity arises for sacrificing himself and showing the glorious negative energy arising from his long course of training in patience, devotion, and subjection, which alone enable him to attain to this high degree of passive courage. His heart is free from fear of death, and with cold disdain the Russian Soldier meets his fate and goes blindly into action, suppressing absolutely all his human feelings. . . .
>
> The Russian dies as calmly and as uncomplainingly as he lives. From his birth this feeling of being tired of life is innate in him, and truly his life is joyless, unattractive, and worthless. As he has neither to trouble for freedom, which he has never known, nor yet for any other of the higher goods of this world, it is not difficult for him to lay down his life amidst the smoke and blood of battle while steadily holding his position, or to let himself be cut to pieces in an attack. Alongside of this fatalism there is no room for the instinct of self-preservation. . . .
>
> Next to steady endurance in the defensive, the lava-like rush on the offensive is the characteristic of the Russian Soldier.[6]

A good example of this "rush" was the Soviet infantry tactic for clearing a minefield during World War II: simply charge through it, sacrificing a few men in order to make a path for their comrades.

Unlike the troops of the army, the common sailors of the navy—like those of all navies—continued to resist unjust discipline. During the general strikes and revolts of 1905, the successful mutiny of the crew of the battleship *Potemkin* at Odessa in the Black Sea signaled more military uprisings over the next two years by no fewer than 15,000 sailors and soldiers. The crews of three destroyers at Vladivostok had to be crushed in 1907 and 14 percent of the personnel of the Baltic Fleet arrested. This spirit of resistance to indifferent treatment contributed greatly to the Bolshevik Revolution in 1917, in which the sailors of Kronstadt mutinied. No Communists, they again revolted four years later against their new masters and had to be crushed.[7]

The Communist state under Lenin and Stalin—like Ivan IV, a "Terrible" in his own right—acted no differently than had its Mongol and tsarist forebears by insisting on absolute obedience to the state for internal and external security. The secret police (N.K.V.D., M.V.D., K.G.B.) replaced the palace Guards and enabled Stalin to execute his bloody purges of 1937–38. The elimination of former tsarist and politically doubtful officers so deepened the fear of the regime by the masses that the latter even went so far as to welcome and assist the Nazi invasion in 1941. The ruthless reprisals against that piece of disloyalty were predictable. And encirclement by the American-NATO-Chinese triad since World War II has only perpetuated Soviet Russian paranoia about an invasion.

Though the Red Army remains a para-police arm of the Kremlin, the Navy remains less reliable. As recently as 1975 the crew of a Russian frigate in the Baltic, led by its shipboard political commissar, mutinied and almost made good its escape before being overtaken by other units. The Red Navy had no powerful advocate until Admiral S. G. Gorshkov, Commander in Chief of the Navy from 1955 to 1986, and his legacy remains yet to be compared with his continental forebears Colbert, Tirpitz, and Yamamoto.

Much has been made by historians of the thesis of historian Robert J. Kerner that the history of Russian foreign policy can be understood as a relentless urge to the sea and warm-water outlets.[8] Thus Imperial Russia steadily drove the Turks from the Crimea in the Black Sea and spent much of the late nineteenth century pushing eastward to the Pacific, pacifying the descendants of the Mongols in Central Asia to establish Port Arthur and Vladivostok as naval bases on the Pacific coast and to dominate China—until running afoul of Japan in 1904–05. The inference is that Russia would eventually and inevitably become a great maritime and naval power, just as the United States ended its westward expansion across North America by achieving major naval and imperial status in the 1890s. The Russian attempt to conquer the East Asian landmass cannot be disputed; it alarmed the British sufficiently to strengthen the defenses of India and

Afghanistan against possible overland attack, to force the Russians away from Korea, and to ally with Japan in 1902. But Russian overseas expansion had little impetus, revealed most dramatically when Russia sold the only overseas possession in its history, Alaska, to the United States in 1867.

Any strategic comparison between Russian and American expansion into the Pacific Ocean is just as false as that between Japan and Great Britain just because these two nations happen to be formed of islands. The United States expanded overland from a maritime or thalassocratic base. That is, the United States was already a liberal, capitalistic society devoted to overseas trade by the time she absorbed California and began to develop San Francisco and other West coast ports from the 1840s. America's merchant marine already rivaled Britain's by then, and she kept going by opening Japan for trade in the 1850s and finally taking possession of Hawaii and other Pacific islands before the end of the century, then connecting her vast oceanic system with the Panama Canal in 1914. The concurrent American occupation of the continental interior was merely the filling in of the vast wastelands between the great trade entrepots of the Atlantic and Pacific seaboards. Russia had no such maritime base before or after she reached the Pacific; her ports remained as icebound as her continental form of government.

The disparate geostrategic outlooks of the United States and Soviet Russia, in the words of Michael MccGwire of the Brookings Institution, have thus been shaped by differing historical experiences: "U.S. attitudes toward the role of projected force have their roots in the European experience of maritime colonialism, founded on naval power. These attitudes were reinforced by World War II. . . . The Soviet Union, meanwhile, had the tsarist experience of continental expansion and a reluctance to take (or retain) possession of distant territories, while their own experience of World War II centered on the ejection and military defeat of the German invaders"—and, it should be added, preventing an attack from the East by keeping some forty front-rank divisions facing the Japanese Kwantung Army in China throughout 1941–45.[9]

Whereas Adolf Hitler in *Mein Kampf* announced his plans to seek *Lebensraum* by the physical conquest of the Russian Ukraine, the professed means of Marxian-Leninist "conquest" are through world revolution, made possible by the internal socioeconomic decay of the industrial West—an arguably fanciful mid-nineteenth century philosophy embraced by the Russian Communists for internal as much as external motives. Militarily, Soviet Russia seeks to maintain its nationalistic fortress strategy while these revolutions spread. The Kremlin has a very long wait, particularly because it is strapped with an economic system historically outclassed by its capitalistic rivals.

ECONOMIC

By its very geographical nature as a sprawling landmass, the continental great power is an agrarian state. The heart of its economy is agriculture, with which it feeds its great numbers of peasants who work the farms and comprise the armies. The landowning aristocracy—and the monarch—thus derives its financial and political power from successful farm production, maintaining local control by feudal rule. The continental power's single-element economy is therefore homogeneous, but also very fragile. For even one poor growing season—or worse, several in succession—can decimate the economy and thereby weaken the state.

The economic pyramid of the continental state comprises only two classes, the small aristocracy and the huge peasantry. No true broad-based bourgeois middle class of merchants, bankers, and shippers exists, for too little agricultural or capital surplus is earned for reinvesting in other profit-making enterprises. Or whatever gold comes in is squandered on luxury items for the privileged few. The economy and the social structure thus lack diversity. Without the manufactures derived from reinvested capital, industry does not exist on a major scale. Manufactured goods must be obtained from abroad, from the maritime trading states—which are usually real or potential enemies.

The measure of success in a continental state is the simple accumulation of wealth for luxury, prestige, and power. So even when a small merchant element exists within the continental state, its goal is to aspire to aristocratic status rather than to remain in the trades. Business is disdained by the men of power, who turn to it only out of necessity—in order to buy land and live as "robber barons." Of course, a monarchy may try to force trade and business upon an unwilling citizenry, but it is an artificial endeavor which lacks any staying power beyond the reign of any such "enlightened" king or queen.

Indeed, the continental superstate is trapped in an economic straitjacket from which there is no escape. Not only is it dependent upon the maritime competitor for manufactured goods, but must even hire foreign bankers, businessmen, and shippers to manage its larger economic affairs. This often includes the production of munitions and ships.

Such a homogeneous and land-bound economic structure has no real need of overseas colonies or trading enclaves. It might send out adventurers, explorers, political agitators, even conquerors to seize enemy or undeveloped regions, but such acquisitions are invariably for political power and prestige and perhaps raw specie (gold and silver) rather than for sustained economic development. Such isolated colonies become exposed to attack or exploitation by rivals, nothing more than useless territorial appendages that drain valuable national treasure for their defense.

The case of France is a perfect example of the continental power which tried to break out of this economic mold and failed. An agrarian society, for centuries its kings and princes ruled in feudalistic seclusion. Only during the reign of Louis XIV in the late seventeenth century did France try to diversify. Louis charged Colbert with creating a merchant marine, navy, and colonial system to rival those of Holland and England. But upon Colbert's death these efforts were quickly subordinated to the defense of the homeland. The colonies contributed little to the economy, except that merchants enriched by the enterprise worked their way into the aristocracy. Gradually, during the eighteenth century, the British drove the French from Canada and India, and when Napoleon rose to power in 1798 he set out for Egypt and India on an orgy of territorial conquests, humble in economic motive. The French "empire" that followed the fall of Napoleon consisted of the utterly insignificant outposts of Algeria, Madagascar, Indochina (Vietnam), and several scattered islands and coastal enclaves. And Napoleon had had to violate his own "Continental System" of economic boycott by trading with the enemy for manufactured necessities.

When the Industrial Revolution reached the continent during the nineteenth century, the landed powers enjoyed greater economic opportunities, made possible by the spread of the Revolution from Britain. That is, the rise of factories and global trade under the aegis of the *Pax Britannica* led to the emergence of a dynamic industrialized merchant class on the continent, especially in Germany. Prussian efficiency turned the new German state into the economic giant of Western Europe, but without transforming the Prussian monarchy and national penchant for authoritarian rule. Kaiser Wilhelm II played at global "empire" by grabbing profitless (and defenseless) chunks of real estate in tropical Africa and the Pacific and encouraged overseas shipping to match his new navy. But in the end, 1914 and 1939, he and Hitler repressed their vibrant capitalistic class to the will of state control, especially under the respective wartime economic czars Walter Rathenau and Albert Speer.

The Japanese economic experience closely paralleled that of Germany, the emerging class of industrialists and shippers—the *Zaibatsu*—being utterly dominated by the ancient baronial samurai class and feudal imperial political system.[10] A country dangerously poor in raw materials, Japan embarked on continental conquest in search of them, beginning with Manchuria in 1931. Japan's new industrialized war machine, however, functioned on oil, obtainable only as imports. When the United States cut this off in 1941, Japan elected to invade the oil-rich Dutch East Indies and rubber-rich British Malaya, which resulted in war with the Western Allies and the ultimate destruction of Imperial Japan's industrial economy.

Though Russia has always had raw materials, until the twentieth century it was a typically agrarian continental state, feudalistic in character in spite of the conscious efforts of Peter the Great to pattern his backward state after Western European nations. Though he drew particularly on Dutch technical advisers to stimulate Russian economic growth, his political and social model was continental France. His greatest need remained feeding the masses of serfs who worked the farms and filled his armies. And the erratic Russian wheat crop has ever since defied the need for economic stability, from Peter's landowning royal successors through the land-sharing commissars. Even the emergent minority merchant class of the late nineteenth century was stifled by a backward "Slavophile" economic conservatism.[11]

From feudalism to communism, Russia has persisted as a continental agricultural society, heavily dependent on imported foodstuffs as well as technological know-how, the exploitation of its vast raw materials being weighted down by state controls. Oddly, Russia's great Marxian political revolution was based on an economic philosophy of industrial workers seizing control of the means of manufacturing to create a pure socialist state. But inasmuch as Imperial Russia never enjoyed a true, powerful middle class of capitalists, the revolution of 1917 amounted to little more than the transfer of political and economic power from the land-owning aristocracy to the land-sharing peasantry on collective farms. This fact gives the revolution the ideological lie, for the Soviet economy remains enfeebled by its failure to promote individual enterprise. Of course, the Kremlin has tried to distract the attention of its subdued citizens by professing to export the "revolution," to the very industrialized nations upon whose imports and technological skills the Russian economy has leaned. In fact, the only appeal of the Russian "world revolution" is in even more economically underdeveloped agrarian lands—the Third World—where the Marxian philosophy is even less viable.

The Soviet state is still fighting the image of the land-owning tsars, a backward, archaic economic determinism doomed to failure in the face of the capitalistic West and its dynamic culture.

CULTURAL

Given the absolute prerequisite for internal security and obedience to the ruler, the closed continental state has invariably repressed free thought. In religion, this has meant a state church, intolerant of rival beliefs, especially those which "protest" the religious dogma in favor of individual fulfillment. The priesthood, furthermore, is a political arm of the state, in pre-industrial times enforcing the theological premise of the divine right of the king. Secular free thinking and foreign influences alike have been similarly rejected as threats to the existing order. The democratic notion of the role of the artist and philosopher as healthy critics of government is anathema, hence new ideas are stifled. And the masses are kept ignorant and obedient through lack of general education or by state-controlled schools.

With creativity, such as it is, constrained to political utility, the culture of the continental state is generally bland and backward, reduced to inferior status in contrast to the dynamic cultural processes of liberal societies. Philosophy, art, science, and technology are subjugated to the will and control of the state. Instead of individual creativity, even in educational institutions, philosophers, artists, scientists, and inventors do their research under the physical eye of political bureaucrats and even the police. Progress in these fields is therefore behind that in the maritime, democratic nations, from which the major advancements and discoveries are imported, imitated, or simply stolen by means of espionage. Even the initial success of the Sputnik satellites was due to captured German rockets and scientists. By the same token, creative individuals, frustrated by such repression, flee to the free societies in order to fulfill their intellectual needs. To counter this, the continental state either forbids their travel beyond its borders, imprisons them as subversive, or, if they are lucky, expels them altogether.

France inherited the mantle of continental leadership from feudal Catholic Spain in the seventeenth century, and though it instituted no formal Inquisition of heretical free thinkers the French monarchy so insisted upon religious orthodoxy that it forced its Protestant Huguenots to flee the country. Culturally behind the glorious worlds of the Dutch and English Renaissance, Louis XIV charged Colbert with importing great minds and creating French artistic and scientific academies. But such thinkers often abandoned France in disgust at the political repression, while native Enlightenment intellectual giants like Voltaire and Descartes endured the Bastille or fled abroad. Great architecture could create a Versailles, but only as opulence for the aristocracy, and the ideas of political reformers like Montesquieu had their major effect on other nations.

Even the French Revolution was a religious crusade, the Committee of Public Safety and subsequent Directory being a veritable church to supplant the nationalistic priests who served royalty. Such chauvinistic (a French word) loyalty was easily transferred to Bonaparte, who, in spite of his despotism, became and remains to this day almost diefied in the French psyche. A distrust and dislike of foreigners, foreign ideas, and even foreign languages prevailed throughout. In spite of the brilliance of such artistry as that of the late nineteenth century Impressionists, their works were rejected by the Academy. And liberal ideas were subordinated to popular prejudices like the pervasive anti-Semitism epitomized by the Dreyfus affair. The promise of high French culture died in World War I.

The modern German state evolved from the Holy Roman Empire, a loose, religiously centered, political symbol that survived the Protestant Reformation, which had been spawned in the Germanies by Martin Luther in the sixteenth century. The energetic Germans developed a dynamic culture over the ensuing three centuries and with overtones of political liberalism, only to have it dashed by the Fatherland Front of Ludendorff in World War I. His Prussianism admired the Japanese warrior code of *Bushido* and was easily transformed into the racial prejudices of Hitler's "master race." Emigrés like Albert Einstein escaped before the Holocaust attempted to exterminate the Jewish race.

High German culture—in art, science, literature, education, psychology, technology—succumbed to the Nazi order, in which even the magnificent music of Wagner was distorted to serve the gangsters of the Third Reich. Even wartime research was centralized under the watchful eyes of the S.S. (*Schutzstaffel*, private army of Nazi thugs) and Gestapo, inhibiting the free experimentation in private laboratories which helped give the Allies the margin of technological victory.

Japanese culture never escaped the straitjacket of Asian feudalism. The emperor was considered semidivine, his absolute will derived from the ancient Shinto religion of ancestor worship. For this, the serene people of Nippon became fanatical in their service to him, willing to die gloriously in battle as a religious duty. The modern Western world was feared and despised by the reactionary culture of Japan, where ideas were simply kept to one's self. For its science and technology, imperial Japan depended utterly on Western models, an imitative backwardness typical of everything modern in Japanese society. Imperial Japan was a cultural anachronism in the modern world until forced to Westernize completely by the American victor.

Imperial Russia, though ethnically and geographically a Western nation, did not evolve naturally from the Middle Ages with the rest of Europe, because of the long Mongol occupation and even after it had been forcibly dragged into the modern world by Peter the Great in the early eighteenth century. Regarding France as his cultural model, Peter overturned medieval custom by adopting French as the second national language, shearing the men folk of their traditional beards, and clothing them in Western European garb. To do this, however, he had to utilize rather than circumvent the powerful Russian Orthodox Church and its priesthood, which therefore remained an arm of the state down to the twentieth century.

The Communist Revolution supplanted church dogma with the unbending secular religion of Marxism—equally rigid in its orthodoxy. The Communist Party became the new priesthood, though carefully wedded to traditional patriotism. "Soviet patriotism," boasts the Red Navy today, "is based on the principle that love for the native land, its language, culture and history, and respect for revolutionary and national traditions are combined with loyalty to the Communist Party, Soviet government, and people. . . . Soviet officers and sailors have a common Marxist-Leninist ideology, and are firmly united by this ideology and their aspiration for the victory of communism. . . . Soviet naval commanders are . . . politically experienced organizers and educators [who] . . . rely on Party and Young Communist League [*Komsomol*] organizations. . . . The forms and methods of educating naval personnel are diverse, and one of the most important is political education. Propaganda through lectures, talks, political briefings, [etc.] . . . plays a significant role."[12]

A resigned fatalism has continuously permeated the peasantry of serfs, aided by escapism through alcoholism, while the great cultural expressions of especially nineteenth-century Russian literature were primarily critical commentaries on this servile condition. In spite of imitating Western manners and external trappings, the state spurned foreign and liberal ideas as internal and external threats, among which were the Moslem Turks, the ideals of the French Revolution, and finally, under the Soviets, the high culture of the democracies.

Eight Centuries of Continental Strategy

Lenin failed to create a proletarian people's culture in the 1920s or even to stifle, for example, the immense appeal of free-wheeling American jazz music. And anti-Semitism has run as a common thread throughout all Russian history.[13]

The Communist regime swiftly arrested the budding intellectual roots of free criticism under the late Imperial period by exiling, imprisoning, or exterminating free-thinking artists, literati, engineers, philosophers, and religions as subversive to the order of the state. Russian technology has thus lacked the qualitative excellence of that of the West. Even through dogged determination, stolen secrets, and outright imitation, Russian technical and scientific accomplishments have lacked the sophistication to prevent a Chernobyl nuclear accident. Russian culture remains conservative and dull, out of touch with the modern world—a sad reality reflected in the daily lives of the people. It is a dreariness made manifest by the fabric of the Russian police state, dominated by the army.[14]

MILITARY

Given the geopolitical facts of the continental state as previously discussed in this essay, the army of every continental great power has been defensive in character. The very borders of the continental state have consequently been regarded as military "frontiers," a bastion of fixed fortifications and permanent armies standing behind them. Indeed, the military history of modern continental Europe has been largely written around frontier quarrels—wars over disputed borderlands for better security such as the Alsace-Lorraine. Continental dictators aimed at outright permanent conquest of neighboring great powers have been the exception rather than the rule in modern history and invariably unsuccessful over the long term. The continental navy has been essentially subsidiary as an army transport and commerce raiding force.

Although seventeenth century France under Louis XIV sought to dominate Europe politically, its military strategy was founded on the great fortifications of Marshal Vauban along the Rhine frontier. The French army never carried more than three days' rations because it was not supposed to operate at any great distance beyond French soil.

French military policy did not change markedly down through the Maginot Line of the early twentieth century, with the singular exception of Napoleon, whose megalomania led him on repeated invasions of conquest throughout Europe. But even his program, an aberration in the continuum of French military doctrine, failed finally in Russia and Spain. The marshals of France from Turenne to Foch reflected and supported the defensive character of the French state. They kept the navy in a subordinate role by having the admirals wage *guerre de course*, commerce warfare, merely to weaken the British economy, rather than engaging the enemy fleet for command of the sea.

The modern Germany army of the Wilhelmian Empire was conservative to the core, having inherited its defensive character from the struggles for survival epitomized by Frederick the Great's Prussia in the Seven Years War of 1756–63. The Moltkes and Schlieffen planned to defeat France by a bold stroke through the Low Countries only to gain territorial concessions along the frontier. When the French slowed the German advance in 1914, the short-ranged German army lost its momentum, and four years of static trench warfare resulted. Like Napoleon, Hitler's megalomania enabled him to over-rule the conservative field marshals of the traditional army then conquer France in 1940 but also to fatally overextend himself and Germany's resources throughout Europe and even North Africa, but especially in Russia. Tirpitz's pre-1914 navy developed as a diplomatic tool to thwart the British navy from risking war in support of France and Russia. But in both world wars the German navy, like France's of old, was devoted to commerce warfare, using surface raiders and U-boats; its fleet dared not challenge the British fleet for command of the sea.

Imperial Japan consciously modeled its army after that of Germany and sought to create an empire centered on the continent of Asia. In the centuries preceding 1900, Japan's armies were confined to the home islands, but the extension of the Russian military presence into the Far East by that year caused Japan to seek the expulsion of the Russians from the lands, islands, and seas facing her homeland. Japan's fanatical generals gradually committed the army to the conquest of East Asia, beginning with victory in the first Sino-Japanese War in 1895, then the Russo-Japanese War in 1905, and finally undertaking the second Sino-Japanese War in 1937 and engaging in Mongolian border clashes with the Russians in 1938 and 1939. The Japanese attempt to carve out a continental empire—the so-called Greater East Asia Co-Prosperity Sphere—brought defeat at the hands of the Chinese Communists over 1937–45 and the Red Army thrust through Manchuria in 1945.[15]

The Imperial Japanese Navy was an inshore fleet committed throughout its short history to supporting the Army on the continent, including the destruction of the enemy fleet in Japanese home waters. This succeeded at Tsushima in 1905 and was to have been realized against the United States, except that Admiral Yamamoto's daring long-distance carrier raid accomplished the purpose at Pearl Harbor in 1941. Unique among continental great powers, Japan rejected commerce warfare against the United States and instead decided to contest the U.S. Pacific Fleet for command of the sea in 1942–45. This decision gradually backfired with the American naval victories from Midway to Okinawa and the ultimate air-sea blockade of the home islands, culminating in the atomic bombs. Japan had dared break out of her strategic continental mold only at her peril—and failed.

The fortress Russia described by Mahan at the beginning of this essay is a more typical military example of the continental state than Japan. Throughout its history Russia has maintained fortified frontiers along its many exposed borders against the Mongols, Teutonic Knights, Turks, Swedes, Napoleon, Germans, Japan, and now China and NATO. As a great flat topographical bowl, Russia has been open to invaders from the Steppes of Asia and the high ground of Central Europe. Hence the determination of the Red Army and the Kremlin never to abandon Central Europe, bought so dearly from the Wehrmacht at the cost of twenty million lives in World War II.

Inasmuch as control over the high ground eluded Imperial Russia, from earliest times the cities were surrounded by defensive walls and the frontiers inhabited by the "Garrison Army." Walls were traditional devices in place of natural topographical barriers like mountains or coastlines. Ivan the Terrible, for example, strengthened his southern frontier against the Tartars at the Oka River south of Moscow in 1572 by floating a River Force and perfecting the "Walking Fort" along the banks: 9′ × 12′ × 1-foot-thick sections of palisaded wooden walls on wheels, linked for miles and guided by the troops. This ingenious defense, the very year after the Tartars had destroyed Moscow, crushed the Tartar attack and killed 100,000 of their troops.

The fact that Ivan was able to field successive armies in prolonged wars could not have been possible without the fatalistic stoicism of the average Russian soldier—unchanging down the centuries. Neither was initiative expected or encouraged among the officers. Obedience to the state was and is the basic tenet of conduct in a continental army on the defensive, a truth discerned by an English merchant living in Moscow in the early sixteenth century: "The Russe trusteth rather to his number than to the valure of his souldiers, or good ordering of his forces. . . . Their footemen (because otherwise they want [lack] order in leading [command]) are commonly placed in some ambush or place of advantage, where they most annoy the enemie with least hurt to themselves. . . . The Russe souldier is thought to be better at his defense within some castle or town, than hee is abroad. . . ."[16]

Two hundred years later the brilliant General Burkhard C. Munnich, a German mercenary in Russian service, established a "magazine" (fort) system in depth along the frontier—a cordon strategy that allowed each man but three days' rations for operations beyond the borders of Russia (the French example). It proved to be too thin but reflected the traditional Russian penchant for defensive strategems. More effective were the scorched-earth policy used against Napoleon in 1812 and the defense of road junctions to slow the Wehrmacht in 1941–42. *And* the deadly Russian winters.

The history of the Russian navy is one of uninterrupted defensive roles, the navy being completely subordinated to the army as a transport and support force along the coasts and in the rivers and lakes. Although the "Ship Fleet" existed from Peter's time, its few line-of-battle ships were rarely committed to open battle. Yet Peter I, Catherine II, and other monarchs shared the fascination for such battleships as awesome symbols of national prestige. In spite of the disaster of Tsushima, Nicholas II authorized more battleships. And Stalin tried to purchase major warships from the United States as part of a naval buildup in the 1930s, though in vain.[17]

Eight Centuries of Continental Strategy

Like its continental forebears, Soviet Russia has since World War II created an impressive navy of major superior weapons and equipment, in quantity if not always quality. And like Britain of old, the United States has responded to the new rival continental "fleet" with alarm. But without overseas bases and allies, Russia is still in no position, geographically, to mount and sustain overseas amphibious operations *of any kind* as long as her seaborne lines of communications are disputed by the U.S. Navy. Furthermore, like the Germans before them, once Russian warships and submarines escape their confined territorial waters, their safe return becomes highly problematical.

Russia is simply unable—and thus unwilling—to wage a prolonged war at sea. This includes any limited war. Sir Julian Corbett wisely observed in 1911 that only a true maritime power could successfully wage a limited war—because with its superior navy, used only against islands or peninsulas, it is able to completely isolate the battlefield from outside supply and reinforcement while its army subdues the enemy at leisure. Russia has *never in her history* been willing or able to attempt such an overseas adventure. Her lines of communication have always been over *land*, contiguous to the borders of Russia, and even then her ability to pacify restive neighbors has been questionable. Witness Afghanistan in the 1980s.

What Winston Churchill so aptly described in 1947 as the Iron Curtain across Central Europe was and is nothing less than the ultimate Russian military frontier to prevent yet another invasion of "Holy Mother Russia"—after perhaps one hundred of them throughout the country's turbulent history. To the Soviet Union, the NATO armies, particularly West Germany's, are merely the same military horse but only of a different color.

The Allied military interventions against the Bolsheviks in 1919 initiated Communist Russian paranoia over the hostile intentions of the Western democracies, thus adding an ideological dimension to the Kremlin's desire to create political buffer zones along its military frontiers. Hence the Warsaw Pact satellites have been kept in line by passive or active Russian arms, accounting for the interventions in Czechoslovakia, Hungary, and Poland since the late 1940s, as well as an active presence in North Korea and the foray into Afghanistan in the 1980s. Russian paranoia over the integrity of its borders even led to the destruction of a large South Korean commercial airliner when it violated Soviet air space in 1983.

Just as the British Empire during the nineteenth century assumed that Imperial Russia was intent on conquering the entire Eurasian landmass, including India, so too has the United States since World War II engaged in similar worst-case thinking vis-à-vis the Soviet Union. By focusing on Russian capabilities rather than intentions, American strategy makers fell into the trap of viewing all Soviet military activity as a "relentless buildup" for an armed conquest of the world. They have pointed to the years 1948 to 1953 for their evidence, the time when the USSR was filling in the strategic vacuum on its exposed borders left by the collapse of the arch-enemy German and Japanese continental empires. American counteraction during these years quickly arrested further expansion of Russia's military frontiers, which have since even contracted by its loss of the Chinese alliance after 1960. In fact, Russian Communism was contained by the end of 1951 and has absorbed no more neighboring countries since then.

"The Soviet Union," MccGwire has observed, "sprawls across half the Eurasian landmass, the traditional enemies on its borders are now all ranged against it, and it has no significant or reliable allies."[18] Small wonder, then, that the Russian state *must* remain an armed camp—just as would the United States were it flanked territorially by a hostile NATO instead of Mexico or a billion-soul China instead of Canada.

Since 1945, therefore, the Soviet Union has built up an impressive arsenal of weapons at every level but without using them. Except for Afghanistan, a hopeless quagmire and drain on Soviet morale, the Red Army and Air Force have tested their equipment in combat only through confederates of the Third World. The Soviet Rocket Forces of course match the American strategic missile deterrent, though both are equally useless in conventional military terms. All the evidence points to extreme reluctance on the part of the Kremlin to risk its survival in a nuclear war of any type. To date, the Soviet surface navy has been little more than a parade fleet, an excellent tool for gunboat diplomacy but yet to be tested—or risked—

in a shooting situation. The navy in fact has been used primarily as a responsive tool to Western naval power projection around the world, never initiating American-style offensive operations—like a blockade of Cuba (1962), an intervention like Grenada (1983), punitive air strikes like against Libya (1986), or an outright sustained overseas limited war like Vietnam (1964–73).

As Fred Jane said in 1899, "Ivan realizes that he exists to be shot at. . . ." The Russian navy, like the army, is still defensive.

CONCLUSION

Given the foregoing historical summary of continental France, Germany, Japan, and Russia, there can be no doubt that definite consistent patterns of strategic behavior have been shared by all four great continental powers. Although slight variations exist, to be sure, particularly in the case of Oriental Japan, these four superpowers of the modern period have exhibited the same general geopolitical, economic, cultural, and military characteristics of behavior.

By contrast, as shown in the preceding essays of this volume, the great maritime powers or thalassocracies—notably, for the modern period, the Netherlands, Great Britain, and the United States—have behaved in an exactly opposite manner. They have been geopolitically island republics, democratic and capitalistic, based upon individual rights and free enterprise, devoid of large *standing* armies (meaning in the home country in peacetime), but dependent upon large standing navies to police the seas, protecting overseas trade, markets, and colonies. And they have allied with continental powers in wartime.

The question which the contemporary Western alliances, dominated by maritime America, must decide is whether the Soviet Union of the late twentieth century has broken from its historical roots and patterns of strategic behavior. Has the Communist system brought significant change to continental Russia? Can Soviet Russia become a maritime state, geopolitically, economically, and culturally as well as militarily? Indeed, has the advent of thermonuclear weapons made Russia any less paranoic about risking annihilation from "invasion" by aerial bombardment? Or, in the conventional sphere, is Russia now inclined to launch an offensive invasion of Western Europe—or of China, or India, or Africa, or the Middle East—*for the first time in her history?*

Contemporary American foreign and defense policy is ill served if based essentially on what might be termed the "Pearl Harbor syndrome." Just because Japan caught us off-guard in her sneak attack of 1941 hardly implies that such a strategy must needs lie at the root of the strategies of all our future potential enemies. By the same token, we must respect the very real and seering fear in the minds of all Russians of the "Hitler syndrome"—the inevitable desire of the Western powers to bludgeon and conquer the Russian homeland as in the struggle of 1941–45. The Russian anxiety over invasion is in fact more legitimate than our own, which stems from the loss of ships and only 2,400 lives on American soil and water against *20,000,000* by the Russians defending their homeland against Hitler! And, indeed, whereas the United States has experienced only one Pearl Harbor, Russia has endured dozens of Hitlers, but most profoundly the Mongols, Charles XII of Sweden, and Napoleon—not to mention the Allied interventions in the Russian Civil War against Bolshevism in 1919.

One may conclude, from the historical evidence, that Russia is now as always committed to the survival of the state (the Soviet regime), the church (the Russian Communist Party), and the homeland (Mother Russia). The Marxian gospel may preach global revolution, bankrupt though it is as a secular theology, just as Americans may desire an entire world made safe for our brand of democracy, also not likely for many reasons—in places like Vietnam, the Middle East, and central Africa.

Beyond these questions and observations, the strategic historican can say little more. He has done his duty by interjecting the historical dimension into the discussion over the possibilities and probabilities of Russian strategic behavior. The final conclusions can only be seen fifty or one hundred years hence, by the historians of the future.

NOTES

1. Fred T. Jane, *The Imperial Russian Navy* (London, 1899), pp. 518–519. Italics mine.

2. Alfred Thayer Mahan, *Naval Strategy* (Boston, 1911), pp. 383–401, reprinted in Allan Westcott, ed., *Mahan on Naval Warfare* (London, 1919), pp. 256–269.
3. Captains 1st Rank G. A. Ammon and M. I. Grigoriev, Rear Admiral Y. A. Grechko, and Colonel N. G. Tsyruhnikov (tr. by Joseph Shapiro), *The Soviet Navy in War and Peace* (Moscow, 1981), p. 146.
4. Mahan, "The Interest of America in International Conditions" (1910), reprinted in Westcott, pp. 302–303.
5. R. W. Daly, "Summation of the Course in Russian Military and Naval Doctrines" (U.S. Naval Academy unpublished manuscript, c. 1966), p. 11.
6. Major Otto Wachs, 1889, quotation courtesy of Daly.
7. An excellent recent treatment of the Kronstadt mutinies is Israel Getzler, *Kronstadt, 1917-1921: The Fate of a Soviet Democracy* (New York, 1983).
8. Robert J. Kerner, *The Urge to the Sea: The Course of Russian History* (Berkeley, 1926).
9. Michael MccGwire in Leon Wofsy, ed., *Before the Point of No Return* (New York, 1986), pp. 110–111.
10. See P. N. Davies, "The Rise of Japan's Modern Shipping Industry," *The Great Circle* (Journal of the Australian Association of Maritime History) (April 1985), pp. 45–56.
11. See Alfred J. Rieber, *Merchants and Entrepreneurs in Imperial Russia* (Chapel Hill, 1982), and Thomas C. Owen, *Capitalism and Politics in Russia: A Social History of the Moscow Merchants, 1855-1905* (New York, 1981), and Herbert E. Meyer, "This Communist Internationale Has a Capitalist Accent," *Fortune* (February 1977), pp. 134–148.
12. Ammon, *Soviet Navy*, pp. 104–118.
13. See Marianna Tax Choldin, *A Fence Around the Empire: Russian Censorship of Western Ideas under the Tsars* (Durham, 1985), and S. Frederick Starr, *Red and Hot: The Fate of Jazz in the Soviet Union, 1917-1980* (New York, 1983).
14. See David K. Shipler, *Russia: Broken Idols, Solemn Dreams* (New York, 1983).
15. Raymond L. Garthoff, "Soviet Operations in the War with Japan, August 1945," *U.S. Naval Institute Proceedings* (May 1966), pp. 50–63.
16. Giles Fletcher (who lived in Moscow 1517–1526), *Of the Russe Common Wealth* (London, 1856).
17. See Donald C. Watt, "Stalin's First Bid for Sea Power, 1933–1941," *U.S. Naval Institute Proceedings* (June 1964), pp. 88–96, and Thomas R. Maddux, "United States-Soviet Naval Relations in the 1930s: The Soviet Union's Efforts to Purchase Naval Vessels," *Naval War College Review* (Fall 1976), pp. 28–37.
18. MccGwire, "Misreading the Kremlin," *World Policy Journal* (Fall 1986), especially pp. 682–683, 688ff.

INDEX

224